Hybrid Systems and Multi-energy Networks for the Future Energy Internet

Hybrid Systems and Multi-energy Networks for the Future Energy Internet

Yu Luo
National Engineering Research Center of Chemical Fertilizer Catalyst (NERC-CFC), Fuzhou University, Fujian, China

Yixiang Shi
Department of Energy and Power Engineering, Tsinghua University, Beijing, China

Ningsheng Cai
Department of Energy and Power Engineering, Tsinghua University, Beijing, China

ELSEVIER

ACADEMIC PRESS

An imprint of Elsevier

Academic Press is an imprint of Elsevier
125 London Wall, London EC2Y 5AS, United Kingdom
525 B Street, Suite 1650, San Diego, CA 92101, United States
50 Hampshire Street, 5th Floor, Cambridge, MA 02139, United States
The Boulevard, Langford Lane, Kidlington, Oxford OX5 1GB, United Kingdom

Notices
Knowledge and best practice in this field are constantly changing. As new research and experience broaden our understanding, changes in research methods, professional practices, or medical treatment may become necessary.

Practitioners and researchers must always rely on their own experience and knowledge in evaluating and using any information, methods, compounds, or experiments described herein. In using such information or methods they should be mindful of their own safety and the safety of others, including parties for whom they have a professional responsibility.

To the fullest extent of the law, neither the Publisher nor the authors, contributors, or editors, assume any liability for any injury and/or damage to persons or property as a matter of products liability, negligence or otherwise, or from any use or operation of any methods, products, instructions, or ideas contained in the material herein.

Library of Congress Cataloging-in-Publication Data
A catalog record for this book is available from the Library of Congress

British Library Cataloguing-in-Publication Data
A catalogue record for this book is available from the British Library

ISBN: 978-0-12-819184-2

For information on all Academic Press publications
visit our website at https://www.elsevier.com/books-and-journals

Publisher: Brian Romer
Acquisitions Editor: Lisa Reading
Editorial Project Manager: Chris Hockaday
Production Project Manager: Prasanna Kalyanaraman
Designer: Victoria Pearson

Typeset by Thomson Digital

Working together
to grow libraries in
developing countries

www.elsevier.com • www.bookaid.org

Contents

6. **Ammonia: a clean and efficient energy carrier for distributed hybrid system**

7. **Power balance and dynamic stability of a distributed hybrid energy system**

8. **Applying information technologies in a hybrid multi-energy system**

Acknowledgments

The authors would like to appreciate the support and contribution of Prof. Lilong Jiang, the director of National Engineering Research Center of Chemical Fertilizer Catalyst (NERC-CFC) on Chapter 6. Prof. Lilong Jiang shared his insightful views, interdisciplinary thinking, and comprehensive vision, which help us open our mind and have a new understanding about the potentials of ammonia (NH_3) as a clean and efficient energy carrier. The authors would also like to appreciate the contribution of Yi Zheng from the Department of Energy and Power Engineering, Tsinghua University on Chapters 8 and 9.

The authors would like to appreciate the support from the Beijing Natural Science Foundation Outstanding Youth Science Foundation Project (JQ18009), the National Youth Talent Support Program, and the National Basic Research Program of China (973 Program, 2014CB249201).

Chapter 1

Introduction

Yu Luo[a], Yixiang Shi[b] and Ningsheng Cai[b]

aNational Engineering Research Center of Chemical Fertilizer Catalyst (NERC-CFC), Fuzhou University, Fujian, China; bDepartment of Energy and Power Engineering, Tsinghua University, Beijing, China

1.1 World energy

Energy is one of the most significant basis of human activities. The history of energy transitions reflects the history of the development of human society. As science and technologies develop rapidly, industrialization implements and the human population booms, an increasing amount of energy is needed to produce food, offer clean potable water, process raw materials, support communications, lighting and mobile devices, etc., in order to satisfy the requirement for comfortable and diversified lifestyles.

The original form of energy is a resources existing in nature, which is called as the primary energy sources. The common primary energy sources include fossil fuels, hydro energy, solar energy, wind energy, nuclear energy, etc. Table 1.1 summarizes the global primary energy consumption by energy source in 2018 and the expected one in 2050 released by Energy Information Administration (EIA) from US Department of Energy [1]. In 2018, global energy structure is still dominated by fossil fuels, that is, coal, natural gas, petroleum, etc. The use of fossil fuels accounts for more than 80% of global primary energy consumption, reaching 5.23×10^{11} GJ. About 15% of global primary energy consumption, that is, 9.8×10^{10} GJ, is attributed to renewable energy, and nuclear energy accounts for less than 5% of global primary energy consumption (3.0×10^{10} GJ). According to the EIA's prediction [1], global primary energy consumption in 2050 will rise by almost a half to 9.53×10^{11} GJ, in which renewable energy will rise rapidly by $\sim$3% per year in the coming decades. In 2050, the global renewable energy consumption is expected to climb up by 2.68 folds to 2.63×10^{11} GJ, exceeding any other primary energy sources. Coal, natural gas, and liquid fossil fuels (represented by petroleum) are expected to rise only by 11%, 43%, and 22% to 1.88×10^{11} GJ, 2.08×10^{11} GJ and 2.54×10^{11} GJ, respectively. Relatively slow increase leads to the shares in total primary energy consumption dropping by 6.3%, 0.3%, and 5.5%. Even so, fossil fuels are still expected to account for almost 70% of global primary

Hybrid Systems and Multi-energy Networks for the Future Energy Internet.
http://dx.doi.org/10.1016/B978-0-12-819184-2.00001-8

TABLE 1.1 Global primary energy consumption in 2018 and their expectation in 2050 [1].

Energy sources	Realized primary energy consumption in 2018		Expected primary energy consumption in 2050	
	Quantity (billion GJ)	Share (%)	Quantity (billion GJ)	Share (%)
World total	651	100	953	100
Renewables	98	15.1	263	27.6
Nuclear	30	4.6	40	4.2
Natural gas	145	22.2	208	21.9
Coal	169	26.0	188	19.7
Petroleum and other liquids	209	32.1	254	26.6

energy consumption in 2050. Overall, carbon-free energy sources are playing a more and more significant role in the future energy consumption, but fossil fuels will still dominate in the global energy structure.

EIA also summarized the global end-use energy consumption by sector and end-use fuel in 2018 and the expected one in 2050 as shown in Table 1.2 [1]. Whenever at present or in the future, industrial sector always accounts for more than a half of end-use energy consumption. Transportation sector slightly exceeds one quarter of global end-use energy consumption. The remaining end-use energy consumption (less than one quarter) is attributed to building sector (residential and commercial). At present, the end-use energy consumption is mainly dominated by fossil fuels involving liquid fuels, natural gas and coal, accounting for over 77% of global end-use energy consumption. Among these fossil fuels, liquid fuels are the main fuel or feedstock of transportation and industrial sectors owing to high energy density, acceptable cost and appropriate chemical properties, accounting for 42% of global end-use energy consumption. Coal is still an indispensable industrial end-use fuel used in energy-intensive industrial processes like metallurgy, cement production, etc. Especially in China, coal is still a major energy form widely used in centralized chemical plants. Meanwhile, most of electricity, heat and cold generation are still relied on large-scale coal-fueled power plants, combined heat and power (CHP) plants and combined cooling, heat, and power (CCHP) plants. Since the Second Industrial Revolution, electricity (secondary energy) has been widely used in the residential and commercial buildings. In end-use side, electricity consumption reached 8.2×10^{10} GJ, 17.4% of global end-use energy consumption. With renewable energy blooming and energy conversion technologies and

TABLE 1.2 End-use energy consumption in 2018 and their expectation in 2050 [1].

Sector	Realized primary energy consumption in 2018		Expected primary energy consumption in 2050	
	Quantity (billion GJ)	Share (%)	Quantity (billion GJ)	Share (%)
World total	472	100	655	100
Industrial	250	53.0	334	51.1
Transportation	127	27.0	175	26.7
Residential	61	12.8	95	14.4
Commercial	34	7.2	51	7.8
End-use energy form	Realized primary energy consumption in 2018		Expected primary energy consumption in 2050	
	Quantity (billion GJ)	Share (%)	Quantity (billion GJ)	Share (%)
World total	472	100	655	100
Renewables	24	5.2	36	5.4
Coal	72	15.2	87	13.4
Natural gas	95	20.1	142	21.7
Electricity	82	17.4	148	22.6
Petroleum and other liquids	199	42.1	242	36.9

information technologies developing, energy systems are transforming in the distributed (flexible-scale) and electrification direction. Electricity is expected to play a more and more significant role in the industrial and transportation sectors. In 2050, electricity is expected to account for 22.6% of global end-use energy consumption, about 5% higher than that in 2018. Despite renewable energy sources only account for ~5% of end-use energy consumption at present or in the coming several decades, electricity produced from renewable energy sources will grow significantly according to EIA's data in Table 1.1.

1.2 Electricity

Table 1.3 reveals the global end-use electricity consumption by sector in 2018, as well as the expected one in 2050 reported by EIA [1]. Currently, global electricity use is estimated to be 22.8 trillion kWh, in which industrial sector consumes 10.4 trillion kWh electricity, dominating the global electricity consumption (accounting for ~46%). Residential sector consumes almost 30% of global

TABLE 1.3 End-use electricity consumption by sector in 2018 and their expectation in 2050 [1].

Sector	Realized electricity consumption in 2018		Expected electricity consumption in 2050	
	Quantity (trillion kWh)	Share (%)	Quantity (trillion kWh)	Share (%)
World total	22.8	100	41.1	100
Industrial	10.4	45.9	14.9	36.2
Transportation	0.5	2.1	2.5	6.0
Residential	6.7	29.3	14.7	35.7
Commercial	5.2	22.7	9.1	22.1

electricity use, that is, 6.7 trillion kWh, and commercial sector consumes ~5.2 trillion kWh, almost 23% of global electricity use. As expected, global electricity use will rise by 80% to 41.1 trillion kWh. Electricity use of industrial sector will rise by 43% to 14.9 trillion kWh. The growth rate of industrial electricity use is lower than that of global electricity use, thus, the share of industrial electricity is expected to drop by 10 points of percentage to 36%. The electricity use from residential sector is expected to soar up remarkably by 2.2 folds to almost 14.7 trillion kWh. The share of residential electricity will reach almost 36%, equivalent to that of industrial electricity. The electricity consumed by commercial sector is expected to rise by 76% to ~9.1 trillion kWh, the share of which slightly drops by 0.6%–22.1%. The electricity used by transportation sector is expected to be almost triple of that in 2018 due to the development of electric vehicles. However, the share of transportation sector in electricity use is expected to be still no more than 6%. In the coming future, much more electricity is needed to meet diverse lifestyles of human beings. Therefore, distributed power generation could increase to meet the end-use electricity demand from residences, commercial buildings, electric vehicles, etc.

1.3 Renewable energy

IEA's reports indicate that in 2010, electricity generation from renewable energy accounted for ~21% of global electricity generation [1], and the share rose to 24% in 2016 [2]. As IEA's 2019 report expects [1], in the period from 2018 to 2050, electricity generated from renewable energy will rise by 3.6% per year in average, faster than electricity generated from any other energy resources. Electricity generation from renewable energy is expected to be ~8.2 trillion kWh in 2020, and increase to ~21.7 trillion kWh in 2050. The ratio of electricity generated from renewable energy to global electricity generation will further

TABLE 1.4 Global electricity generation from various renewable energy sources in 2020 and 2050 expected by EIA [1].

Renewable energy sources	Electricity generation in 2020		Electricity generation in 2050	
	Quantity (trillion kWh)	Share (%)	Quantity (trillion kWh)	Share (%)
Renewables	8.16	100	21.72	100
Hydroelectric	4.70	57.6	6.02	27.7
Wind	1.76	21.6	6.74	31.0
Solar	1.29	15.8	8.31	38.3
Geothermal	0.11	1.3	0.24	1.1
Other	0.30	3.7	0.41	1.9

rise to ~31% in 2020, and soar up to ~49% in 2050. Table 1.4 shows the expected electricity generation from various renewable energy sources including hydroenergy, wind energy, solar energy, geothermal energy, etc, in 2020 and 2050 [1]. Until 2020, electricity generated from hydroenergy still dominates renewable energy-driven electricity generation, and wind energy and solar energy will share ~22% and ~16% of global renewable energy used for electricity generation, respectively. Nevertheless, in 2050, renewable energy-driven electricity generation is expected to be dominated by solar energy, wind energy and hydroenergy, in which hydroenergy is expected to share less than solar and wind energy in global renewable energy used for electricity generation. Solar energy is expected to be the largest renewable energy sources used for electricity generation, that is, ~38%. The share of wind energy could be slightly lower than that of solar energy, ~31%. Almost 70% of renewable energy-driven electricity generation is expected to come from solar energy and wind energy. Considering the features of intermittency and fluctuation, the wide application of solar energy and wind energy needs scale-flexible and operation-flexible energy storage systems to balance renewable energy output and end-use electricity demand. For efficient electricity generation or co-generations, distributed generation systems are also necessary. Therefore, distributed hybrid energy systems are one of the essential technologies for large-scale utilization of renewable energy sources.

1.4   Carbon dioxide emission

Current fossil fuel-based energy structure results that carbon dioxide emissions are bound to an inevitable issue. According to the statistical data and expected data from BP [3], Fig. 1.1 shows annual CO_2 emissions by sector in 1995 and

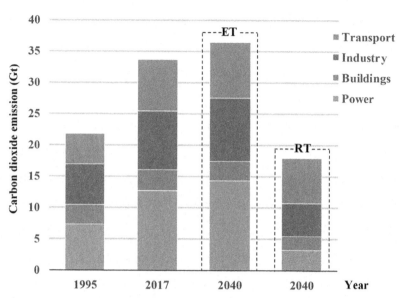

FIGURE 1.1 CO_2 emissions by sector and their 2040 expectation in ET and RT scenarios. *(Data from BP Energy Outlook: 2019 edition [3].)*

2017, and their expectation in 2040. From 1995 to 2017, annual CO_2 emissions increased from 22 Gt by 54% to 34 Gt. This increase should be mainly attributed to power and transport sectors, which have caused that CO_2 emission rose by 5.4 Gt and 3.3 Gt, respectively. BP also presented two scenarios to predict CO_2 emissions in 2040 expectation, named as evolving transition (ET) scenario and rapid transition (RT) scenario, respectively. In the ET scenario, population, gross domestic product (GDP), energy use, policies, technologies, etc., evolves in a manner and speed seen over the recent past [3]. The RT scenario reveals the most optimistic prediction, where all the low-carbon scenarios in all the sectors have been considered to minimize carbon dioxide emissions. In the ET scenario, annual total carbon emission in 2040 is ~7% higher than that in 2017, revealing a much slower growth rate than that before 2017. However, this prediction is still far away from the realization of Paris climate goals. Therefore, much more efforts should be paid on CO_2 emission reduction. Among all the sectors, power sector was expected to be the biggest contributor of CO_2 emission in the coming several decades. In the RT scenario, the annual total CO_2 emission was expected to be almost a half of that in 2017, which is mainly attributed to significant reduction in power sector (reducing by 75%) and industrial sector (reducing by 75%).

To promote CO_2 emission reduction, we should pay our efforts on three aspects, that is, resource efficiency improvement, development of low-carbon energy sources and carriers, and carbon storage and removal [3]. The crucial point for CO_2 emission reduction is to utilize low-carbon even zero-carbon energy sources and develop decarbonized power. On one hand, decarbonized power needs more renewable power, nuclear power, or power generated from fossil fuel-based power plants with carbon capture, utilization, and storage (CCUS). On the other hand, novel clean energy carriers like hydrogen and bioenergy are suitable for meeting energy demand from end-use sides. Meanwhile, efficiency enhancement in energy systems can also help to reduce CO_2 emission. To approach to the goals in RT scenario, renewable power is needed to scale up, and advanced low-carbon technologies are in urgent demand.

1.5 Summary

To develop low-carbon and energy-saving energy roadmap, a series of supporting technologies are required, such as energy storage technology, demand side response technology, grid interconnections and multi-energy networks, energy use electrification, and digitalization. As renewable energy is playing a more and more significant role in the energy structure, energy systems tend to be more scale-flexible and decentralized. Therefore, these supporting technologies will be combined to develop novel distributed hybrid systems, and further promote the formation of the future *Energy Internet*.

In this book, we aim to introduce the concept of the distributed hybrid systems, review a part of novel technologies applicable in the distributed hybrid systems, and share our understandings on the future combination between energy technologies and information technologies. This book consists of 10 chapters. In this chapter (Chapter 1), we briefly reviewed current energy status and the expected energy structure in the coming decades from EIA and BP. In Chapter 2, we will present a brief review on the status of the distributed hybrid systems and supporting technologies. In Chapter 3–7 promising low-carbon energy carriers and their related energy conversion devices, as well as system layout, typical performance and system dynamics of advanced distributed hybrid systems. In Chapter 8 and 9, we will extend to energy networks and discuss their combination with advanced information technologies, especially the artificial intelligence (AI) technologies, and share our perspective on the future *Energy Internet*. Finally, we will outlook the opportunities and challenges for the distributed hybrid systems and the future "Energy Internet."

References

[1] DOE-EIA. International Energy Outlook 2019 with projections to 2050. 2019. <https://www.eia.gov/outlooks/ieo/pdf/ieo2019.pdf>.

[2] T.M. Gür, Review of electrical energy storage technologies, materials and systems: challenges and prospects for large-scale grid storage, Energ Environ Sci 11 (2018) 2696–2767.

[3] BP. BP Energy Outlook: 2019 edition. <https://www.bp.com/content/dam/bp/business-sites/en/global/corporate/pdfs/energy-economics/energy-outlook/bp-energy-outlook-2019.pdf>.

Chapter 2

Distributed hybrid system and prospect of the future Energy Internet

Yu Luo[a], Yixiang Shi[b] and Ningsheng Cai[b]

[a]National Engineering Research Center of Chemical Fertilizer Catalyst (NERC-CFC), Fuzhou University, Fujian, China; [b]Department of Energy and Power Engineering, Tsinghua University, Beijing, China

2.1 Introduction

Increasing energy demand and environmental issues urge human beings to develop and utilize energy sources in a more sustainable pathway. Traditional fossil fuel-based energy structure needs to be optimized by integrating clean energy sources such as renewable energy. Traditional fossil fuel-based centralized energy system (CES) generates electricity using several large-scale generation units, and then transmits electricity to domestic, commercial, and industrial consumers [1]. In CES, the energy devices are generally with large capacities ($\sim$100 MW) and the energy flows are unidirectional [2]. However, CES requires high stability and reliable power integration, which conflicts with the intermittency and fluctuation of renewable energy including wind energy, solar energy and more, and further limits its penetration. To alleviate this conflict and utilize various energy sources adequately, developing distributed energy systems (DESs) is a feasible approach. Fig. 2.1 compares the schematics of centralized and DESs [1]. Differently from large-scale CESs, DESs have the following features:

1. User-oriented: DES serves local energy consumers and directly orients to the local energy demands. DES usually locates near the users, which reduces energy losses and economic costs during the transportation.
2. Flexible scale: The size of DES is flexible and up to the energy consumers. Typically, DES can vary from $\sim$20 W (residential-use generators) to $\sim$10 MW (biomass generation) [3].
3. Multienergy input and output: Development of energy conversion technology and human society leads to the diversification of energy sources and demands. DES is an open energy system, and capable of various energy

Hybrid Systems and Multi-energy Networks for the Future Energy Internet.
http://dx.doi.org/10.1016/B978-0-12-819184-2.00002-X

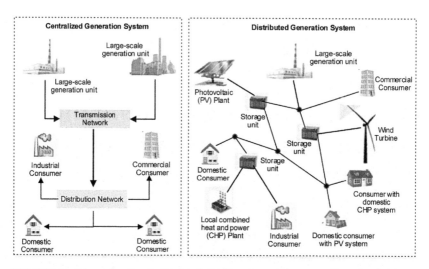

FIGURE 2.1 **Schematic of centralized and distributed energy systems.** *(From Ref. [1]. Copyright 2017 Elsevier Ltd.)*

input and output to utilize energy sources more efficiently and meet diverse energy demands. Bi-directional energy flows are also available in DES.

4. Ability in individual customization: Diverse consumer demands could expand the functions of DES by combining various technologies from multiple fields such as energy engineering, information engineering, automotive engineering, chemical engineering, etc. The evaluation criterion of DES is varying with the demand on efficiency, reliability, economic costs, environmental impacts, and sustainability.

Based on the above features, DES can achieve energy cascade utilization, feasible sizing, and operation to improve the overall efficiency and lower the environmental pollution, particularly for co-generation of electricity, heat, and cold [4,5]. Fig. 2.2 shows the statistic or planned annual installed capacity, share of capacity additions and investments in DES in 2000, 2012, and 2020 [1]. The

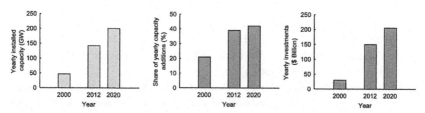

FIGURE 2.2 **Annual installed capacity, share of capacity additions and investments in DES worldwide.** *(From Ref. [1]. Copyright 2017 Elsevier Ltd.)*

newly installed DES capacity soared from 47 GW by 202% to 142 GW, the share of capacity additions from 21% to 38%, and the corresponding investments from 30 billion dollars by 4 folds to 150 billion dollars in the worldwide over the period from 2000 to 2012. As projected, DES will continue growing. The newly installed DES capacity in 2020 will reach 200 GW, 41% of total annual capacity additions, and the annual investments will increase to 206 billion dollars.

Long distance electricity transmission is made feasible by increasing voltage, thus decreasing transportation losses [6]. However, heat and cold are transported through a certain working medium like water. Heat with higher temperature or cold with lower temperature results in higher thermal losses and lower exergy efficiency during the transportation. The use of water as the transport medium of thermal energy is limited in temperature range of $<300\,°C$ and transport distance of <10 km [7]. Consequently, DES can avoid the issue in long distance transmission of heat and cold and feasibly select the device capacity according to local energy demands. Besides, centralized plants aim to meeting the common energy demands for most consumers. DES is designed according to the oriented users, hence, can specifically consider some specific requirements. Particularly, off-grid DES can provide a feasible option to solve the power supply in rural or other remote areas where the power grid is hard to cover. Moreover, faster response of DES leads to a more flexible control and regulation, hence, DES can decouple intermittent renewable energy and power grid to enhance the local utilization of renewable energy. Furthermore, some advanced but scale-limited energy devices such as fuel cells, electrolyzers, etc. have been available in DES, which significantly improves the performance of DES.

Development of electrical, control, and information technologies enables us to look ahead to the future DES. In the next stage of DES, distributed systems will be able to connect with each other to form distributed energy networks (DENs) [8]. DENs will combine the advantages of DES and information networks, hence, offer a new perspective for efficient multienergy generation and supply, reliable and stable operation, as well as economical energy trading. From the new perspective, the concept "Energy Internet" (EI) was proposed. Energy Internet was first proposed by Jeremy Rifkin in *The Third Industrial Revolution* [9]. In the future Energy Internet, energy utilization will be highly distributed and energy devices will be plug-and-play, which enables energy consumers to transmit energy flexibly like information [10]. In Rifkin's opinion, the future Energy Internet will include super-large-scale distributed generation and energy storage systems making it possible to utilize renewable energy as the primary energy widely [9]. Consequently, the future Energy Internet is expected to be the product of DES and DEN.

In this chapter, we introduce the basic topology, applications, and current status of DES and DEN. Furthermore, we look ahead to the trend of DES and discuss the prospect of the future Energy Internet.

2.2 Topology of distributed hybrid systems

In the early stage of DES, the systems are designed to meet the loads of single energy form, respectively. Currently, DES is generally used for co-generation of multiple energy such as combined heat and power generation (CHP), combined cold, heat and power generation (CCHP) systems. In addition, the integration of intermittent renewable energy sources and the diversity of end-use loads also propose new requirements for DES. This requires DES to integrate various energy devices into a distributed hybrid system (DHS) to meet more functions such as fluctuation forecasting, stability and reliability improvement, efficiency enhancement, multienergy synergy management, etc. Conventionally, researchers design a DHS by optimizing power generation at fixed loads. This approach has made a great progress in energy cascade utilization [4,5,11–13]. In a real user-oriented DHS, energy demands are varying with time, especially at a small scale. Thus, a topology of DHS should consider the features of user loads. Energy consumption and energy demand should be two interrelated indicators in a DHS. When integrated into a DHS, intermittent renewable energy sources increase the mismatch between energy supply and demand. DHS also needs energy storage to shift this mismatch in a certain time period. Besides, there is a certain energy remaining unexploited in DHS, which could be recovered from the perspective of energy cascade utilization [14–16].

In consequence, an advanced and high-efficiency DHS consists of four subsystems: energy generation subsystem, energy storage subsystem, energy recovery subsystem, and energy end-use subsystem, as Fig. 2.3 shows.

2.2.1 Energy generation subsystem

Energy generation subsystem generally includes the main energy devices to generate electricity, heat or cold in the DHS. Typically, the energy devices in energy generation subsystem only allow unidirectional energy flows.

2.2.1.1 Fossil fuel-based energy generation

Conventional energy devices for fossil fuels utilization include gas turbine, steam turbine, internal combustion engine and burners. Fuel cells are also optional for power generation from fossil fuels.

Fig. 2.4 shows the efficiency of these optional energy generation technologies based on "simple" fuels such as natural gas or refined liquid fuels at different power rating [17]. Conventional energy devices generally belong to heat engines limited by Carnot efficiency. The efficiencies of advanced turbine system and gas turbine combined cycle can reach over 50%, but the rated power is required to be higher than 100 MW. A 100 kW-level gas turbine or internal combustion engine only has efficiency of 30% even lower. In DES (<10 MW), efficiency of heat engines significantly drops with decreasing power rating. Fuel cells are electrochemical devices with the efficiency beyond Carnot efficiency, hence,

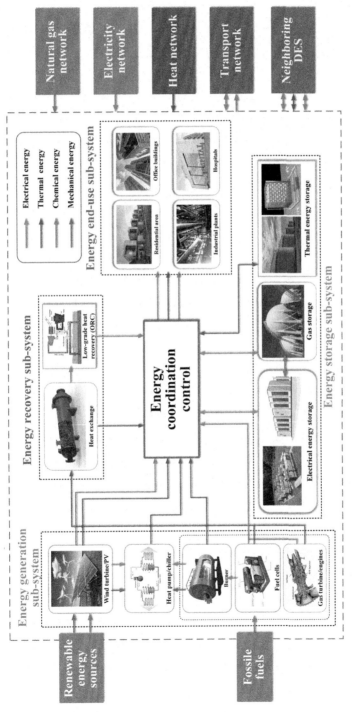

FIGURE 2.3   Schematic of a typical distributed hybrid system.

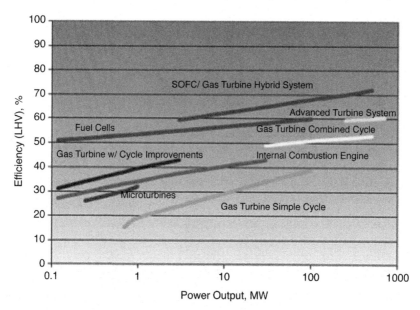

FIGURE 2.4 Efficiency of various fossil fuel-based energy generation technologies at different power rating. *(From Ref. [17]. Copyright 2010 Elsevier Ltd.)*

can still maintain high efficiency (>50%) even at a small scale (~100 kW). Furthermore, advanced high-temperature fuel cells such as solid oxide fuel cells (SOFCs) are possible to be integrated with gas turbine to achieve energy cascade utilization and efficiency improvement in energy generation subsystems [17–20]. In a distributed system, high-grade heat can be generated by burning fuels through burners, or upgraded from low-grade heat through heat pumps. Cold can be generated from compression refrigeration such as air conditioner or adsorption refrigeration (typically using LiBr as absorbent).

2.2.1.2 Renewable energy generators

Clean renewable energy sources can reduce the utilization of fossil fuels to alleviate the environmental impacts of DHS. When DHS utilizes renewable energy sources including solar energy, wind energy, biomass and hydro-energy, energy generation subsystem should integrate solar photovoltaic (PV), wind turbine, biomass generator, and micro-hydro turbine, respectively. Table 2.1 summarizes the contributions of these devices for utilizing different renewable energy sources from the perspective of technique, economics, and environment [1,21–23]. Biomass generator has similar contributions with gas turbines or engines. Technically, biomass generator has the contributions of grid reinforcement, energy loss reduction, reliability and power quality improvement, and supply security, while the other three renewable energy utilization devices can only improve power supply security. The inappropriate placement and sizing of

TABLE 2.1 Contributions of different renewable energy utilization technologies [1].

Devices	Technical contributions		Economic contributions		Environmental contributions	
	Power supply security	Dispatch ability	Transmission operating costs	Fossil fuel costs	Green-house gas emissions	Noise pollution
Solar PV	↑	—	↓	↓	↓	↓
Wind turbine	↑	—	↓	↓	↓	—
Biomass generator	↑	↑	↓	↓	↓	—
Micro-hydro turbine	↑	—	↓	↓	↓	—

renewable energy utilization devices can even result in voltage instability and power quality degradation. Economically, these four renewable energy utilization devices can reduce the operating costs for transmission and distribution, the emission costs and fossil fuel costs. Environmentally, replacing a part of fossil fuels with renewable energy sources helps to reduce greenhouse gas emissions and conserve natural resources.

2.2.2   Energy storage subsystem

Energy storage subsystem aims to balance the mismatch between energy supplies and demands in different caused by the variability and uncertainty of renewable energy resources and end-use loads. The variations in DHS differ by time scale, hence, requires different energy storage characteristics [24,25]. Beaudin et al. [25] summarizes the main functions of energy storage subsystems as shown in Table 2.2. In a microscale time period, the integration of intermittent renewable energy sources requires energy storage devices with fast response to smooth the variation in power generation, increase regulation reserves and control frequency depending on the power system characteristics. In a mesoscale time period, the variation in energy generation or end-use demand may require higher operating reserves to balance energy supply and demand [26], such as peak shaving, renewable capacity firming, and economic optimization. For this purpose, DHS requires approximately 8% of the total power capacity [27]. A case study on Ontario energy system revealed that integrating 10 GW wind power into the Ontario system with peak loads of 26 GW would increase regulation requirements by 11% and operating reserves by 47% [25]. In a longer

TABLE 2.2 Functions of energy storage and required characteristics [25].

Function	Values ($/kW)	Storage time at rated capacity (h)	Power requirement (MW)	Required response time
Renewable capacity firming	172–323	1–10	—	Minutes
Renewable contractual time of production payments	655	4–10	—	Minutes
Smoothing microscale wind variations	—	0.01–0.05	—	<1 cycle
Smoothing mesoscale wind variations	—	0.33–2	0–10	<1 cycle
Smoothing macroscale wind variations	—	2–3000	0.1–1	Seconds–minutes
Removing effects of intermittent clouds on PV	—	0.33	Up to PV capacity	<1 cycle
Distributed generation peaking	—	1	0.5–5	<1 min
Voltage and frequency control, and governor controlled generation	25	0.005	—	<1 cycle
Electric service power quality	717	0.0027–0.017	—	<1/4 cycle
Peak shaving	600–832	2–10	>100	Minutes
Regulation control	789–2300	0.16–1	175–600	—
Bulk electricity price arbitrage	200–300	1–10	—	—
Central generation capacity	215–753	2–6	—	—
Transmission support	169	0.00056–0.0014	1–100+	<1/4 cycle
Reducing transmission access requirements	72	1–6	—	—
Transmission congestion relief	20–72	2–6	—	Minutes

Distribution upgrade deferral	666–1067	2–6	—	Minutes
Transmission upgrade deferral	650–1200	1–6	10–100+	Minutes
Time-of-use energy cost management	1000–1650	4–6	—	Minutes
Demand charge management	465–1076	1–11	<1	<1 min
End-user electricity service reliability	359	0.08–5	<1	<1/4 cycle
Spinning reserves	72–258	0.05–4	1–400	<3 s
Standing reserves	72	1–5	1–100	<10 min
Minimization of trade penalties	—	2	—	Minutes
Telecommunications back-up	—	2	0.001–0.002	<1 cycle
Uninterruptible power supply	—	2	0.05–2	Seconds
Load following	—	>2	1–100+	<1 cycle
Emergency back-up	—	24	1	Seconds–minutes
Seasonal storage	—	>168	50–300	Minutes

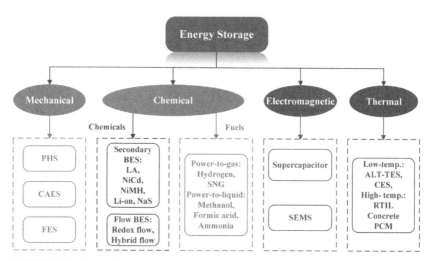

FIGURE 2.5 **Alternative energy storage technologies.** *(Modified from Ref. [30]. Copyright 2014 Elsevier Ltd.)*

time period, the dramatic and unpredictable variation in DHS will impact unit commitment and scheduling of conventional generators, which may require seasonal storage or power supply. Even in some emergencies such as power failure of energy generation subsystem, energy storage subsystem can operate as backup power to follow all the end-use power loads. In this case, DHS needs approximately 15%–20% of the annual loads to meet 2–3 month of storage [28,29].

To satisfy the functions at various time scale, researchers have developed a series of energy storage technologies as Fig. 2.5 shows [30]. Currently, the existing energy storage technologies allow to store energy in the forms of mechanical energy, chemical energy, electromagnetic energy, thermal energy, etc. Table 2.3 compares the technical features and capital costs of these alternative energy storage technologies. Next, we briefly introduce the principles and features of different energy storage technologies.

2.2.2.1 Mechanical energy storage

Storing energy in mechanical energy is relatively mature among these energy storage technologies. Typical energy storage technologies include pumped hydro storage (PHS), compressed air energy storage (CAES), and flywheel energy storage (FES).

PHS has the electrical energy storage technology with the largest scale and highest technical maturity as Table 2.3 and Fig. 2.6 shows. Nearly 99% of the installed electrical storage capacity, i.e., over 120 GW, is PHS. The principle of PHS is to pump water into the reservoir at higher elevation for electrical energy storage, and release water in the higher reservoir to drive hydro turbines for electrical energy generation. Table 2.3 indicates that the PHS has a relatively

TABLE 2.3 Technical features and capital costs for different energy storage technologies [30].

Storage technology	Rating		Density		Response time	Round-trip efficiency (%)	Self-discharge per day (%)
	Power rating (MW)	Discharge time	Power density (W L^{-1})	Energy density (W h L^{-1})			
PHS	100–5000	1–24 h+	0.1–0.2	0.2–2	min	70–80	Very small
CAES	5–300	1–24 h+	0.2–0.6	2–6	min	41–75	Small
FES	0–0.25	s–h	5000	20–80	<s	80–90	100
LA	0–20	s–h	90–700	50–80	<s	75–90	0.1–0.3
NiCd	0–40	s–h	75–700	15–80	<s	60–80	0.2–0.6
Li-ion	0–0.1	min–h	1300–10,000	200–400	<s	65–75	0.1–0.3
NaS	0.05–8	s–h	120–160	15–300	<s	70–85	20
VRB	0.03–3	s–10 h	0.5–2	20–70	s	60–75	Small
ZnBr	0.05–2	s–10 h	1–25	65	s	65–75	Small
PtG	0–50	s–24 h+	0.2–20	600 (200 bar)	s–min	34–44	0
SC	0–0.3	ms–1 h	40,000–120,000	10–20	<s	85–98	20–40
SMES	0.1–10	ms–8 s	2600	6	<s	75–80	10–15

(Continued)

TABLE 2.3 Technical features and capital costs for different energy storage technologies [30] *(Cont.)*

Storage technology	Lifetime		Capital cost		
	In years	In cycles	$ Kw^{-1}	$ (kW h)$^{-1}$	$ (kW h)$^{-1}$ per cycle
PHS	>50	>15,000	600–2000	5–100	0.1–1.4
CAES	>25	>10,000	400–8000	2–50	2–4
FES	15–20	104–107	250–350	1000–5000	3–25
LA	3–15	250–1500	300–600	200–400	20–100
NiCd	5–20	1500–3000	500–1500	800–1500	20–100
Li-ion	5–100	600–1200	1200–4000	600–2500	15–100
NaS	10–15	2500–4500	1000–3000	300–500	8–20
VRB	5–20	>10,000	600–1500	150–1000	5–80
ZnBr	5–10	1000–3650	700–2500	150–1000	5–80
PtG	10–30	103–104	>10,000	—	6000–20,000
SC	4–12	104–105	100–300	300–2000	2–20
SMES	—	—	200–300	1000–10,000	—

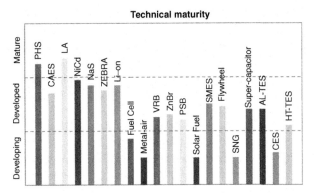

FIGURE 2.6   Technical maturity of alternative energy storage technologies. *(Modified from Ref. [30]. Copyright 2014 Elsevier Ltd.)*

high efficiency (70%–80%), long lifetime (>50 years), large discharging time (>100 MW), and negligible discharging losses at a low cycle cost ($0.1–1.4 per kW h per cycle). These advantages make PHS applicable for grid reliability and wind variability application, particularly for meeting the need of seasonal energy storage [24,31]. However, PHS has a relatively slow response and low flexibility in scale and geographical conditions. Installing PHS will affect the local environment by flooding 10–20 km² of land and require long project lead time (~10 years). These features limit the application of PHS in medium- or small-scale DHS.

CAES is to utilize surplus electricity to compress air and store it underground or in a vessels or pipes at 4–8 MPa [27]. The compressed air can mix with natural gas to generate power through a modified gas turbine. Table 2.3 reveals that CAES has high power rating (5–300 MW), long lifetime (>25 years), and small self-discharge losses. Therefore, CAES is another option for energy storage over seasons even over a year. The capital cost of CAES is $2–4 per kW h per cycle, only higher than that of PHS. Energy storage using flywheels has excellent cycle stability, fast response time, little maintenance cost, high power density (~5000 W L⁻¹), and high round-trip efficiency (80%–90%), hence, FES is capable of suppressing fast renewable power fluctuation and improving power quality. The short discharge time (seconds–hours) and high self-discharge (100% per day) limit the stand-alone application of FES, but it is feasible to combine FES with other energy storage technologies to be a hybrid energy storage subsystem [32].

2.2.2.2   Chemical energy storage

Storing energy in the form of chemical energy offers relatively wide optional energy storage technologies due to the diversity of chemical storage medium. Generally, current energy storage technologies mainly store energy in batteries and fuels.

Battery energy storage (BES) technologies can be classified to secondary battery energy storage and flow battery energy storage. Lead acid battery (LA), nickel cadmium battery (NiCd), nickel-metal hydride battery (NiMH), lithium-ion battery (Li-ion), and sodium sulfur battery (NaS) are all alternative technologies for secondary batteries. As Table 2.3 shows, secondary batteries has fast dynamic response, high power density and high round-trip efficiency, hence, are suitable for the application of load follow to enhance the stability of DHS. Besides, secondary batteries has the advantages of short lead time, potential convenient sitting and technology modularity, which can promote its construction in DHS [27,30]. Nevertheless, metal toxicity is a significant ecological problem for secondary batteries. Among various secondary batteries, LA batteries have relatively higher round-trip efficiency but lower lifetime in cycles. The limitation of operating cycles remarkably reduces the capital cost per cycle of LA batteries. NiCd batteries have higher lifetime cycles, enabling its application in large-scale energy storage (as high as 40 MW). However, the lifetime of NiCd batteries could significantly reduce when suffering deep cycles, and their operation is also affected by the memory effect [30]. Li-ion batteries perform better in power and energy densities. Li-ion batteries are widely studies in their application of electrical vehicles. The capital cost per kW of Li-ion batteries is relatively high, and this limits the application scale of Li-ion batteries. NaS batteries operating at 300–340°C show the highest lifetime cycles and the lowest capital cost per cycle, but the highest self-discharge per day, thus, NaS batteries are more economical in the applications of frequent charge-discharge processes such as power quality regulation. Compared with secondary batteries, flow batteries exhibit longer discharge time and much lower self-discharge losses due to external storage of liquid electrolyte. Vanadium redox battery (VRB), polysulfide bromide (PSB), and zinc bromine battery (ZnBr) are typical flow batteries. Particularly, the lifetime cycle number of VRB reaches over 10,000, which reveals its potential in large-scale DHS. Furthermore, Fig. 2.6 shows LA batteries have the highest technical maturity, and the other BES technologies are technically developed and commercially available. As a whole, secondary batteries have higher technical maturity than flow batteries.

Storing energy in fuels is another feasible chemical energy storage technology. Surplus energy can be converted into gaseous fuels such as hydrogen and synthesis natural gas (SNG), etc., and liquid fuels such as methanol, formic acid, and ammonia. Conventionally, we can convert surplus or unstable energy to the target product through mature chemical industrial processes, and convert it back to stable energy needed by end-users through heat engines or other thermal equipment. Development in electrochemical, photochemical and photoelectrochemical conversion technologies offers a novel option for direct and bi-directional conversion between renewable energy sources and fuels. For example, power-to-gas (PtG) technology can convert surplus renewable power into gaseous fuels through electrolyzers, and then store gaseous fuels without zero self-discharge losses. Hydrogen is a typical product of PtG. Water electrolyzer can directly generate high purity H_2 when driven by electrical power. When needed, stable electrical power can be generated by H_2 fuel cells or H_2-fed heat engines. When the electrolyzer and fuel cell are combined into one reactor, a complete charge-discharge process can be realized in one energy device, i.e., reversible fuel cell (RFC). RFC is similar with flow batteries. The reactants flow in and the products flow out the cell, but the difference is that the electrolyte of RFC doesn't flow and only transfers ions. As Table 2.3 shows, hydrogen storage using PtG has apparent advantages and disadvantages. PtG has the advantages of high energy density (600 W h L^{-1} at 200 bar), high storage period and zero self-discharge. Other than PHS and CAES, storing energy in fuels is the only energy storage technology applicable for seasonal energy storage [25,28]. Nevertheless, PtG has a relatively low round-trip efficiency especially when using low-temperature electrolyzers and fuel cells. Low-temperature electrolyzers and fuel cells result in the electricity used for overcoming polarization losses in the form of low-grade heat, which make it hard to recover this part of losses. Reversible solid oxide fuel cells (RSOFCs) typically operate at 600–800°C. Higher operating temperature can reduce polarization losses and upgrade the generated heat. When RSOFC integrates with efficient energy recovery subsystem, the overall round-trip efficiency would reach 70%–80% [33,34]. From the perspective of technical maturity and capital cost, PtG using electrolyzers and fuel cells is still in the early stage of commercialization, but quickly developing. Great progress has been achieved worldwide in the recent years. The capital cost is expected to continue decreasing in the future. Other than hydrogen, methane, i.e., SNG, is also an alternative gaseous storage fuel for PtG. Power-to-CH_4 not only offers a better solution for gas storage, i.e., the existing natural gas network, but also recycles and reuse the exhausted carbon dioxide emitted during the fossil fuel utilization [35,36]. Furthermore, storing energy in liquid fuels (power-to-liquid, PtL) such as methanol, formic acid, liquid ammonia, etc., is a considerable and economically feasible energy storage technology due to high energy density and convenience in transmission. However, this approach is only capable of utilizing surplus energy, but has little effects on the improvement in performance and stability of DHS. When distributed chemical industries are integrated in the future DEN, PtL would play a much more significant role in the hybrid networks.

2.2.2.3   Electromagnetic energy storage

Supercapacitor (SC) and superconducting magnetic energy storage (SMES) are two optional technologies to store energy in the form of electromagnetic energy. SC has a much larger capacitance, higher efficiency, longer lifetime, and higher energy density than conventional capacitors [25,27]. The unique advantages of SC are extremely high cycle stability and power density (40,000–120,000 W L^{-1}). Extraordinarily low inner resistance leads to fast charging and discharging of SC. Therefore, SC is applicable for renewable power smoothing in microscale variation and power leveling. Particularly, the battery-supercapacitor hybrid energy storage can cope with power charging/discharging in macroscale period and peak power surges in microscale period [37]. SMES has extremely fast rapid response due to the coil in the superconducting state, thus, can supply active and reactive power. SMES is available for small-scale commercial use, which has a great potential to load leveling and power stabilization in DHS [38].

2.2.2.4 Thermal energy storage

According to operating temperature, thermal energy storage (TES) can be classified into low-temperature TES and high-temperature TES. Heat/cold stored in TES can either supply to heat engine cycles for electricity generation, or directly supply to end-user. Thermal energy loads can be categorized into industrial cooling loads (<18°C), building cooling loads (at 0–12°C), building heating loads (at 25–50°C), and industrial heating loads (>175°C) [27]. TES with operating temperature below room temperature belongs to low-temperature TES, including auriferous low-temperature TES (ALT-TES) and cryogenic energy storage (CES) [27]. ALT-TES is technically developed as Fig. 2.6 shows, and commercially applicable for shaving building or industrial peak cooling loads during the daytime, particularly in large buildings cold. CES is still at developing stage. On the contrary, high-temperature TES operates above room temperature. Molten salt storage and room temperature ionic liquids (RTILs), concrete storage and phase change materials (PCM) belongs to high-temperature TES in use and under development [27].

2.2.3 Energy recovery subsystem

Generally, energy generation subsystem exhausts heat/cold with different grades and unexploited fuels more or less. In energy storage subsystem and energy end-use subsystem, there are still valuable latent and sensible heat exhausted. Particularly in most heat engine cycles, efficient exhaust heat utilization is the key for efficiency improvement. Therefore, energy recovery subsystem is an indispensable part in an advanced DHS to realize better energy cascade utilization. Unreacted low-concentration fuels can be used for upgrading heat by an after-burner, which is an extremely common and mature technology. Here, we only introduce the heat recovery technologies.

2.2.3.1 Heat and energy exchangers

Heat exchanger is a common heat recovery device to recover the sensible heat from hot fluid to heat cold fluid. To recover both latent and sensible heat, energy exchangers are proposed and under development. Table 2.4 lists heat exchange and energy exchange technologies in use and under development [39]. Flat plate heat exchangers, heat wheels and heat pipes belongs to adjacent-duct heat exchange technologies, while run-around glycol loops and thermosiphon are non-adjacent-duct ones. Flat plate heat exchangers and heat wheels are both commercially available. Heat exchangers are also widely applied in DHS integrating heat engine cycles or high-temperature fuel cells [5,11,15,18]. The effectiveness

TABLE 2.4 Classification of heat and energy exchange technologies [39].

Heat exchange—adjacent duct	Heat exchange—nonadjacent duct
• Flat plate heat exchanger • Heat wheel • Heat pipe	• Thermosiphon • Run-around glycol loop
Energy exchange—adjacent duct	**Energy exchange—nonadjacent duct**
• Membrane energy exchanger • Enthalpy wheel • VENTIREG exchanger	• Run-around membrane energy exchanger • Twin tower enthalpy recovery loop

of a flat plat heat exchanger is 70%–90% in counter-flow mode and 60%–80% in cross-flow mode [39], however, current technologies generally adopt cross flow due to simple construction and easy sealing. The effectiveness of heat wheels strongly depend on the flow rates and heat transfer surface area. Heat pipes can realize fast and long-distance heat transfer with small temperature difference through capillarity of fluid, but their effectiveness are relatively low (0.35–0.6) [40]. The application of heat pipes has expanded to the field of fuel cell technologies for better thermal integration and easier thermal management [41,42]. Thermosiphon tubes are similar with heat pipes, but depend on gravity of fluid instead of capillarity and require higher temperature difference [39].

Energy exchanger technologies are relatively newer technologies and generally modified from heat exchangers in order to utilize both latent and sensible heat better. Adjacent-duce energy exchangers include energy wheels, flat plate membrane energy exchangers, VENTIREG exchangers, while nonadjacent-duct energy exchangers include run-around membrane energy exchangers and twin tower enthalpy recovery loop. Table 2.5 compares the sensible and latent effectiveness and other properties of different heat/energy exchange technologies [39]. VENTIREG exchangers can exchange both heat and moisture between exhaust and supply to reduce heat losses and avoid ice formation, particularly in a cold climate location. VENTIREG exchangers can recover up to 95% energy and 70%–90% of moisture, and reach high effectiveness in both the sensible heat (85%) and latent heat (60%). But the application of VENTIREG exchanger is limited in toxic environments. The heat pipes have lower sensible effectiveness. Flat plate membrane operating in quasi counter flow has relatively high sensible effectiveness but relatively low latent effectiveness.

2.2.3.2 Low-grade heat-to-power technologies

Several advanced energy recovery technologies, i.e., organic Rankine cycles (ORCs) [15,43–45] and thermoelectric generators (TEGs) [46–50], can recover low-grade heat to generate electrical energy. Heat-to-power using ORC typically recovers low-grade heat ($<350°C$) to maximize net power output in DHS, which

TABLE 2.5 Comparison of different heat and energy exchange technologies [39].

Type of exchanger	Sensible effectiveness	Latent effectiveness	Moving parts	Use in toxic environments
Flat plate heat exchanger, cross flow	60%–80%	—	No	Yes
Flat plate heat exchanger, counter flow	70%–90%	—	No	If good sealing
Flat plate membrane, quasi counter flow	80%–85%	46%–76%	No	Yes
Heat wheels	50%–80%	—	Yes	No
Energy wheels	50%–85%	50%–85%	Yes	No
VENTIREG exchangers	85%	60%	Slow rotation	No
Heat pipes	35%–60%	—	No	—
Run-around heat exchangers	65%–70%	—	No	Yes
Run-around membrane energy exchangers	60%–80%	50%–65%	No	Yes

is applicable for waste heat recovery and low-grade renewable thermal sources utilization such as geothermal energy in DHS. ORC has a relatively small and flexible scale (50 kW$_e$ to 50 MW$_e$ [45]), high efficiency, simple system configuration and high reliability [51]. Considering the condensation of acidic flue gas, ORC has a cooling limit above the dew point temperature (100–130°C) [45]. A series of architectures for ORC are available, however, a trade-off among performance, system complexity and techno-economic considerations should be made. TEG, as another heat-to-power device, can recover low-grade waste-heat in a temperature range of 25–225°C on the basis of Seebeck effect and Peltier effects observed from semiconductor materials [49]. Compared with other waste heat recovery technologies, TEG has the advantages of silence, small size, higher scalability, and durability [52]. Bi$_2$Te$_3$ is considered as the most effective commercially available TEG material. Different temperature from hot and cold junctions leads to a potential difference between two junctions, and further form current. Therefore, it's the key for power generation improvement of TEG to form a larger temperature difference. The integration of TEG with photovoltaic thermal (PV/T) is a promising pathway to utilize high-temperature heat of solar energy [50]. The small size of TEG makes it possible for the automotive application. BMW, Ford, Renault and Honda all developed systems using TEG for exhaust heat recovery [52]. Generally, TEG is limited to place on the exhaust heat source surface. A novel compact integration of TEG and heat pipes enable to a more flexible design for the energy recovery subsystem in DHS [52].

2.2.4   Energy end-use subsystem

Energy end-use subsystem mainly includes various loads such as electrical loads, heat demands, cold loads, gas loads, etc. The diversity of DHS end-user results in diverse variation of these loads with times. Conventionally, energy efficiency and operating costs are both evaluated at fixed energy loads, i.e., so-called rated loads. However, this static design fails to reveal the inefficiency resulted from the mismatch of energy supply and demand. Actual operating conditions are usually far away from the design operating conditions, which could lead to a deviated system efficiency and extra operating costs. Furthermore, high diversification and individuation are also the trend for the future energy demand. This requires a much higher flexibility of DHS in energy demand forecasting even regulation.

2.2.4.1   Energy demand forecasting

To match energy supply with energy demand in real time, reliable energy demand forecasting is requisite. Advanced information processing technology helps to improve the accuracy of energy demand forecasting. Energy demand forecasting can be classified into long-term (1–10 year ahead), medium-term (1 month to 1 year ahead), and short-term forecasting according to time interval [53], and into deterministic forecasting and data-driven forecasting according to forecasting approach [54]. Long-term forecasting aims at long-term system planning from the perspective of energy policy, medium-term forecasting aims at efficient and economical operation and maintenance of DHS, and short-term forecasting aims at reliable and stable operation, control of spinning reserve and price evaluation. Short-term forecasting is the major concern in existing literature [53] and this chapter. Deterministic forecasting is based on the identified science of physical environment, hence, needs no historical data accumulation for training and is easy to generalize. However, a number of factors such as the energy behavior, occupancy schedule, thermal properties of building materials, weather conditions, individuation of end-user, etc., could affect the real-time energy loads. These factors all increase the complexities and uncertainties of energy loads, leading to the decrease of forecasting accuracy. As for data-driven forecasting, it is based on the machine learning techniques. Data-driven forecasting can offer higher accuracy and fast computation for real-time data after trained by plenty of historical data. Nevertheless, a new untrained model limits its use at the beginning, and slight variation in energy end-user or surrounding environment could result the inapplicability of the trained model.

Data-driven load forecasting is a significant application of artificial intelligence (AI) technologies. Major machine learning techniques for data-driven forecasting include artificial neural network (ANN), autoregressive integrated moving average (ARIMA), support vector machines (SVM), fuzzy time series, grey prediction model, as well as moving average and exponential smoothing (MA&ES), etc. [54]. Table 2.6 shows the application, length of training data, accuracy and calculation time of these six major data-driven real-time forecasting

TABLE 2.6 Application, length of training data, accuracy and calculation time of six major and other hybrid data-driven forecasting [54].

Technique	Applied data	Length of training data	Accuracy MAPE (%)	Calculation time
Single technique with comparable input data				
ANN	Hourly electrical load	1 years	1.69–1.81	35 s
ARIMA	Monthly peak electrical demand	6 years	1.05–2.59	—
SVM	Hourly cooling load + climate data	1 month	1.001–1.016	<1 min
Fuzzy	Daily electrical load	6 months	1.23–1.63	—
Grey	Half-hourly cooling load	4 months	0.416–1.097	—
MA&ES	10–30 min electrical load	7 months	1.2	—
Hybrid techniques				
ARIMA + ANN	Hourly electrical load	4 weeks	0.016	—
Fuzzy + ANN	Monthly or annual data	10–64 years	0.50–6.43	—
Fuzzy + ARIMA	Hourly load + weather data	1 years	2.58	—
ARIMA + Evolutionary algorithms (EA)	Hourly data for 4 seasons + temperature data for 3 cities	1–2 months	2.27–3.42	—
Fuzzy + EA	Half-hourly load	1–2 years	0.79–0.94	—
Fuzzy + SVM + EA	Monthly electricity load	20 years	2.13	—
Grey + EA	Daily load	—	<2	—

with comparable input data and some hybrid data-driven real-time forecasting cases. The collection frequency and length of training data both affect the training results. As a whole, these 6 data-driven forecasting techniques have comparable accuracy in mean absolute percent error (MAPE).

2.2.4.2   Demand response programs

In conventional sense, energy loads in a DHS are absolutely up to energy consumers. We generally believe that DHS aims to follow energy loads and meet the demand of energy consumers. Based on this target, we focus on various advanced energy storage, control, and electric technologies for better load follow, i.e., matching energy supply and demand. Increasing penetration of advanced metering infrastructure and growing application of communication infrastructure and smart electrical appliances may change our concept on system design and management. It is becoming possible to change unidirectional load-follow energy supply pattern to novel bi-directional supply-demand interaction pattern, as known as demand response programs (DRP). As we have realized, many of our energy usage habits are not so efficient. DRP offers a novel pattern to improve system efficiency, enhance system stability and energy supply quality in DHS by affecting the energy consumption behavior of end-user in a dispatchable or nondispatchable pathway. Fig. 2.7 shows currently available schemes for DRP. In dispatchable schemes, DHS with the permission of end-users is allowed to control certain electric appliances for optimizing energy usage or matching with power supply, particularly during the peak or emergency period [55]. The nondispatchable schemes are time-sensitive or price-based, including time-of-use, critical peak pricing, and real-time pricing. These schemes aim to

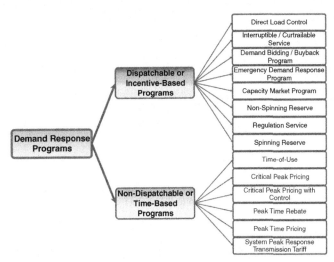

FIGURE 2.7   Classification of demand response schemes. *(From Ref. [55]. Copyright 2015 Elsevier Ltd.)*

flatten energy loads and shave the peak loads through price variation. From the perspective of end-users, load curtailment can be traded in spot market as operating reserves to form a new transaction mode [55].

2.2.5 Connection and interaction

Energy generation, energy storage, energy recovery, and energy end-use subsystems are integrated into a complete and efficient DHS. Of course, most DHS till contain only two or three of these four subsystems, and the missing parts can be supplemented by connecting with neighboring DHS or district energy transmission networks (such as power grid, fuel transmission network, heat network, etc.). A DHS combines renewable energy sources (solar energy, wind energy, hydro-energy, biomass, etc.) and fossil fuels (coal, natural gas, petroleum, etc.) to supply electricity, heat, cold and gas, and also exchanges energy with its neighboring DHS or district energy transmission networks. Consequently, multienergy flows including electrical energy flow, thermal energy flow and chemical energy flow (typically natural gas flow) are formed in DHS, between the neighboring DHS or between DHS and energy transmission networks. Different energy flows have different response features, tolerable time delay and match degree between supply and demand. To deal with such a complicated multienergy supply, DHS require the corresponding control strategies for meeting adequate energy regulation and allocation, stabilizing voltage and frequency, guaranteeing safe switch-over between islanded mode and grid-connected mode, as well as optimizing energy exchange between DHS and DHS or DHS and district energy transmission networks.

2.3 Scales of distributed hybrid systems

DHS can flexibly design its scale to meet different energy demands, utilize different energy sources and achieve different potential functions according to application scenes. Various application scenes also offer different prospects for development of DHS. Elizabeth et al. [3] summarizes the projected DHS at various scales, as well as their benefits and drawbacks as Table 2.7 shows. DHS can be classified into grid-connected DHS, micro-grid-type DHS and islanded DHS.

2.3.1 Grid-connected DHS

Large-scale DHS with installed capacity of 100 kW to tens of MW generally aims to supplement energy supply for power grid or energy-intensive industries. Large power plants are time-consuming, high-cost, time delays, and with political risk. Compared with large power plants, relatively small-scale hydropower unit and wind farm have shorter construction period, hence, reducing investment risks. Besides, some industries with agricultural waste (biomass) like

TABLE 2.7 Classification of various distributed hybrid systems by scale [3].

Scale and type	Typical application scene	Benefits	Drawbacks
5–60 MW, Grid-connected	• Hydro-turbines on a small river • A wind farm • Large diesel generators • Burning agricultural byproducts at a sugar or corn processing facility	• Increase supply; • Leverage investor or multilateral funding; • Limited grid extension required; • Provide income to generating facilities	• Limited application in remote areas far from grid, thus hard to address electricity poverty for the extreme rural poor
<5 MW, Grid-connected	• Rooftop solar • Commercial diesel generators • Wind or hydro with lower capacity	• Increase supply • Back-up electricity to households and businesses • No need to extend grids • provide income to generating facilities	• Limited application in remote areas far from grid, thus hard to address electricity poverty for the extreme rural poor
100 kW to 5 MW+, Micro-grids	• Village households connected to a biomass or hydro plant • Generation system for a business or hospital with surplus generation sold to others locally	• Suitable where grid extension is infeasible, particularly where a business big enough to anchor the system exists • Provide electricity to an entire community	• High cost per kWh • Long-term operability requires an entity (such as a business, village council, or community group) to manage project
60 W to kW-level, Community-use building	• Solar arrays for schools or hospitals • Biomass- or hydro-powered community agricultural processing facilities	• Directly provide development benefits • Projects often attract donor funding	• Difficult long-term maintenance • Do not provide household electricity
20 W to kW-level, Residential-use building	• Solar home systems with solar panel and battery storage • Single wind turbine or small hydro plant	• Household can purchase generation to match its needs and ability to pay	• Costs high relative to incomes • Limited finance options in many places • Relies on end-user maintenance • Supply is limited and may be intermittent

sugar or corn processing factories can also produce electricity through biomass generators. These projects can attract a number of investors and funding, and have been extensively launched in India, Kenya, Uganda, Mauritius, Ecuador, etc. [3].

With the trend of energy utilization miniaturization, the installation of DHS in resident or commercial buildings has been encouraged by some countries as a novel supplement for power grid, as well as a back-up power source in the case of power failure. For example, Germany uses these DHS to promote the utilization of renewable energy sources. Malaysia also encourages to integrate PV into commercial building and allows the owners to sell excess electricity generated from these DHS [56].

2.3.2 Micro-grid DHS

Some DHS aim to effectively utilize local renewable energy resources such as solar irradiance, wind or water flow and share with nearby energy consumers. In a certain area, several DHS connect with each other to form a micro-grid. Generally, the projected DHS in a micro-grid has installed capacity of <5 MW [3]. A micro-grid DHS has much more flexible operation. It can operate in either islanded mode or grid-connected mode. Micro-grid DHS is suitable for somewhere with difficulties in centralized grid extension in short- to medium term. Ref. [3] reveals that micro-grid DHS have supplied electrical power for thousands of rural communities in Asian. Most of these DHS use biomass generator fueled with local agricultural waste due to lower cost. An increasing number of DHS rely on solar energy or wind energy because of their rapid development and policy support. These DHS generally integrate diesel generators in order to reliable electricity supply. A micro-grid requires a specific community to manage and maintain these DHS [3]. For example, a town factories or rural hospitals is quite suitable because of guaranteed basic energy loads and plenty administrative capacity. Somewhere with dense residential area is also favored to reduce costs for energy distribution and infrastructure. Besides, government policies could also be beneficial for the formation of specific social communities.

2.3.3 Islanded DHS

Islanded DHS are applied for a community building or household at a scale of kW-level or lower. For larger-scale applications, typical islanded DHS are built to meet specific energy demands of community buildings like a hospital, clinic, school, community center or factory. These DHS are generally implemented by local business or nongovernment organizations, and typically power only one building instead of wide connection with individual households [3]. Potentially, these DHS could also supply the residential building or rural households where the energy consumers could be unable to afford home-scale DHS. In smaller scale, DHS can be installed at home, typically using solar PV panels on the rooftop owing to their dropping prices year by year. More and more companies in the world wide are engaged in designing small-scale domestic DHS. Some projects also use wind, micro-hydro or natural gas-based DHS. These DHS are flexible in scale, thus, particularly works well in the region with low population and meet their local energy demands such as somewhere remote without power grids or micro-grids. Nevertheless, these projects may suffer risks in low pay-back and high rate of technical failure because of insufficient technical maturity. In Japan, Osaka Gas has been engaged in developing residential CHP system since 2003. Three generation of CHP systems have been commercialized with a scale of 0.7–3 kW: the 1st-generation gas-engine-type system "ECOWILL" in 2003, the 2nd-generation polymer electrolyte fuel cell (PEFC)-type system "ENE-FARM" in 2009, and the 3rd-generation SOFC-type system "ENE-FARM type S" in 2012 [57]. 160,000 fuel cell-based CHP units have been sold until the end of 2016, and this amount is expected to keep increasing to 1,400,000 by 2020 and 5,300,000 by 2030 [58]. The payback time considering unit price and installation cost is expected to be 7–8 years by 2020 and 5 years by 2030.

2.4   Distributed energy networks

Connections among different DHS and district energy transmission networks form a DENs. As Fig. 2.3 shows, connection with energy transmission networks improve the flexibility of DHS and guarantee the stability of distributed energy supply. On the contrary, multienergy DHS is the convertor for different forms of energy, offering an opportunity for the combination of power grid, fuel transmission network (typically natural gas network), district heat network, even electric vehicle(EV)-based transport network. Conventionally, DES aims to the conversion from fuels to electricity, heat and cold. Electricity-to-heat or cold can be realized using electric heater, heat pumps, and chillers. Advanced ORC and TEG also make heat/cold-to-electricity possible. Besides, power-to-gas technology provides a new option to generate hydrogen or methane for natural gas network, which bridges a bi-direction conversion between electricity and gas. Furthermore, apart from meeting local energy demand, intermittent renewable energy sources can be also converted into electrical, thermal or chemical energy for these energy networks. As a result, development of DHS greatly enhances the interaction among various energy networks. This also turns the role of DEN from simple energy generator to energy manager. DEN needs to consider energy generation, energy transport, energy storage and energy allocation to comprehensively optimize energy efficiency, operation reliability and stability, as well as economic feasibility in a wider area. These complicated processes in DEN all significantly increase the difficulty in management and control of DEN. Therefore, hierarchical control should be developed.

More concretely, the functions of hierarchical control from the perspective of electrical power-based DEN are summarized in Fig. 2.8A. Stability,

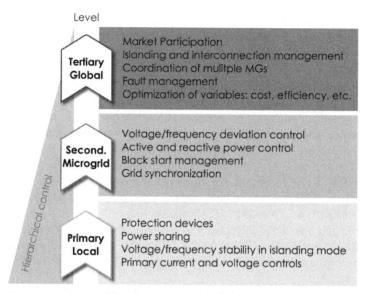

Level

Tertiary
Global
- Market Participation
- Islanding and interconnection management
- Coordination of mulitple MGs
- Fault management
- Optimization of variables: cost, efficiency, etc.

Second.
Microgrid
- Voltage/frequency deviation control
- Active and reactive power control
- Black start management
- Grid synchronization

Primary
Local
- Protection devices
- Power sharing
- Voltage/frequency stability in islanding mode
- Primary current and voltage controls

Hierarchical control

FIGURE 2.8 **(A) Targets of hierarchical control; (B) hierarchical control strategies.** *(Modified from Ref. [59]. Copyright 2015 Elsevier Ltd.)*

protection, power balance, transition, power transmission, synchronization, and comprehensive optimization need to realized synchronously through primary control, secondary control, and tertiary control in DEN. Here, we extend the functions of hierarchical control strategies to multienergy-based DEN. In device level, the primary control implemented by device controllers should realize device protection, stable energy generation and sharing as well as safe switch-over between islanded and network-connected mode. In DHS level, the secondary control needs to balance the deviation between energy supply and demand, manage system self-start-up, as well as effective energy exchange with neighboring DHS, grid, as well as district heat and fuel transmission networks. In the whole DEN level, tertiary control should consider market participation, coordination of multiple DHS, fault management as well as global optimization of DEN, and the global interaction among DHS and energy networks should be also well addressed. To achieve these functions, a hierarchical control strategies for electrical energy-based DEN (micro-grids) are classified in Fig. 2.8B.

2.5 Prospect of the future Energy Internet

When various DEN interact with each other and allow the integration of multiple energy sources, energy can be effectively shared in a wider area to form a large-scale and DEN. A district DEN has formed a prototype for the Energy Internet. Energy router is considered to be the key component for the construction of the future Energy Internet, which is designed to realize energy interaction, energy regulation, and energy flow control among energy networks [60].

Therefore, DHS is not only the basic unit of DEN, but also the base of the future Energy router. In other words, the combination of advanced DHS and hierarchical control forms the future Energy router. The application of advanced energy devices, comprehensive control strategies, advanced information communication and processing technology will enable current DHS to become miniaturized, intelligent and individual. As expected, a smart DHS will be also the basic unit of the future Energy Internet.

The future Energy Internet has four main features [9]:

1. Renewable energy sources are major primary energy sources.
2. A large number of DHSs is allowed to be integrated.
3. Energy share in a wide area can be realized based on the information Internet technologies.
4. Transport network electrification is supported.

Dong et al. [61] describes that the core of the future Energy Internet is the electrical power network, the information Internet and other advanced information technologies are significant basics, distributed renewable energy DHS is basic unit. Dong believes that the future Energy Internet is a complicated multiflow system coupling a number of distributed renewable energy DHS, electrical power network, natural gas network, transport network and other energy transmission systems or networks. The multiple flows consist of energy flows, information flows and mass flows. Particularly, EV-based transport network is more than a simple energy-load network. EVs can become mobile energy storage devices, and actively participate in smoothing power fluctuation and peak shaving. As the miniaturization of DHS, household DHS will be generalized, making it possible to turn homes from passive energy consumers into intelligent, active energy storage and service providers for the future Energy Internet [8]. The switch-over of energy trading pattern from conventional production-to-customer (P2C) to peer-to-peer (P2P) offer a disruptive solution to managing energy supply and demand, and connects every home into a transactive and interactive network. This enables a better global coordination to achieve the goals of energy-efficiency, manage demand and strengthen power grid. However, there are lots of challenges existing to realize the future Internet Energy, including data security and privacy, complex control, expensive energy storage, as well as the lack of energy device interoperability.

2.6 Summary

In this chapter, we introduced the basic topology, applications, and current status of DHS. We divided DHS into four subsystems: energy generation subsystem, energy storage subsystem, energy recovery subsystem, and energy end-use subsystem. Four subsystems form a supply-demand interactive and energy-efficient hybrid system. For each subsystem, we reviewed current energy devices commercially available or under research and development. The application of advanced energy device improves efficiency, stability and reliability of DHS, and also enriches the functions of DHS. DHS has flexible scale, generally varies from tens of watts to tens of megawatts. According to applications of DHS, DHS can be classified into grid-connected DHS, micro-grid DHS, and islanded DHS.

The connection between DHS and DHS or DHS and energy transmission networks forms a complicated DEN. DEN, as a prototype of the future Energy Internet, needs hierarchical control strategies to meet different requirements in device-level, DHS-level and DEN-level.

Finally, we look ahead to the trend of DHS and discuss the prospect of the future Energy Internet. The future Energy Internet is a complicated multi-flow system coupling a number of distributed renewable energy DHS, electrical power network, natural gas network, transport network and other energy transmission systems or networks. Renewable energy sources are expected to be the major primary energy sources. The application of advanced energy devices, comprehensive control strategies, advanced information communication and processing technology will enable current DHS to become miniaturized, intelligent, and individual. This makes it possible to turn homes from passive energy consumers into intelligent, active energy storage and service providers for the future Energy Internet, and further enables a better global coordination to achieve the goals of energy-efficiency, manage demand and strengthen power grid. However, there are lots of challenges existing to realize the future Internet Energy, such as data security and privacy, complex control, expensive energy storage, as well as the lack of energy device interoperability, etc.

References

[1] A. Ehsan, Q. Yang, Optimal integration and planning of renewable distributed generation in the power distribution networks: a review of analytical techniques, Appl Energ 210 (2018) 44–59.

[2] K.G. Di Santo, E. Kanashiro, S.G. Di Santo, M.A. Saidel, A review on smart grids and experiences in Brazil, Renew Sustain Energy Rev 52 (2015) 1072–1082.

[3] E. Baldwin, J.N. Brass, S. Carley, L.M. MacLean, Electrification and rural development: issues of scale in distributed generation, Wiley Interdiscip Rev: Energy Environ 4 (2015) 196–211.

[4] H. Jin, H. Hong, B. Wang, R. Lin, A new principle of synthetic cascade utilization of chemical energy and physical energy, Sci China Ser E: Technol Sci 48 (2005) 163–179.

[5] D. Xu, Q. Liu, J. Lei, H. Jin, Performance of a combined cooling heating and power system with mid-and-low temperature solar thermal energy and methanol decomposition integration, Energ Convers Manage 102 (2015) 17–25.

[6] K.C. Kavvadias, S. Quoilin, Exploiting waste heat potential by long distance heat transmission: design considerations and techno-economic assessment, Appl Energ 216 (2018) 452–465.

[7] Q. Ma, L. Luo, R.Z. Wang, G. Sauce, A review on transportation of heat energy over long distance: exploratory development, Renew Sustain Energy Rev 13 (2009) 1532–1540.

[8] N. Wang, Transactive control for connected homes and neighbourhoods, Nat Energy (2018).

[9] J. Rifkin, The third industrial revolution: how lateral power is transforming energy, the economy, and the world, Palgrave MacMillan, New York, (2011).

[10] A.Q. Huang, M.L. Crow, G.T. Heydt, J.P. Zheng, S.J. Dale, The future renewable electric energy delivery and management (FREEDM) system: the Energy Internet, Proc IEEE 99 (2011) 133–148.

[11] B. Su, W. Han, H. Jin, Proposal and assessment of a novel integrated CCHP system with biogas steam reforming using solar energy, Appl Energ 206 (2017) 1–11.

[12] C. Zou, Y. Zhang, Q. Falcoz, P. Neveu, C. Zhang, W. Shu, et al. Design and optimization of a high-temperature cavity receiver for a solar energy cascade utilization system, Renew Energ 103 (2017) 478–489.

[13] L. Liu, T. Zhu, Y. Pan, H. Wang, Multiple energy complementation based on distributed energy systems—case study of Chongming county, China, Appl Energ 192 (2017) 329–336.

[14] X. Song, L. Liu, T. Zhu, T. Zhang, Z. Wu, Comparative analysis on operation strategies of CCHP system with cool thermal storage for a data center, Appl Therm Eng 108 (2016) 680–688.

[15] T. Zhang, T. Zhu, W. An, X. Song, L. Liu, H. Liu, Unsteady analysis of a bottoming Organic Rankine Cycle for exhaust heat recovery from an Internal Combustion Engine using Monte Carlo simulation, Energ Convers Manage 124 (2016) 357–368.

[16] A. Colmenar-Santos, G. Zarzuelo-Puch, D. Borge-Diez, C. García-Diéguez, Thermodynamic and exergoeconomic analysis of energy recovery system of biogas from a wastewater treatment plant and use in a Stirling engine, Renew Energ 88 (2016) 171–184.

[17] A.F. Ghoniem, Needs, resources and climate change: clean and efficient conversion technologies, Prog Energ Combust 37 (2011) 15–51.

[18] C. Bao, Y. Shi, C. Li, N. Cai, Q. Su, Multi-level simulation platform of SOFC–GT hybrid generation system, Int J Hydrogen Energ 35 (2010) 2894–2899.

[19] C. Bao, Y. Shi, E. Croiset, C. Li, N. Cai, A multi-level simulation platform of natural gas internal reforming solid oxide fuel cell–gas turbine hybrid generation system: Part I. Solid oxide fuel cell model library, J Power Sources 195 (2010) 4871–4892.

[20] C. Bao, N. Cai, E. Croiset, A multi-level simulation platform of natural gas internal reforming solid oxide fuel cell–gas turbine hybrid generation system—Part II. Balancing units model library and system simulation, J Power Sources 196 (2011) 8424–8434.

[21] W.L. Theo, J.S. Lim, W.S. Ho, H. Hashim, C.T. Lee, Review of distributed generation (DG) system planning and optimisation techniques: comparison of numerical and mathematical modelling methods, Renew Sustain Energy Rev 67 (2017) 531–573.

[22] Z. Abdmouleh, A. Gastli, L. Ben-Brahim, M. Haouari, N.A. Al-Emadi, Review of optimization techniques applied for the integration of distributed generation from renewable energy sources, Renew Energ 113 (2017) 266–280.

[23] H.A.M. Pesaran, P.D. Huy, V.K. Ramachandaramurthy, A review of the optimal allocation of distributed generation: objectives, constraints, methods, and algorithms, Renew Sustain Energy Rev 75 (2017) 293–312.

[24] B. Parsons, M. Milligan, B. Zavadil, D. Brooks, B. Kirby, K. Dragoon, et al. Grid impacts of wind power: a summary of recent studies in the United States, Wind Energy 7 (2004) 87–108.

[25] M. Beaudin, H. Zareipour, A. Schellenberglabe, W. Rosehart, Energy storage for mitigating the variability of renewable electricity sources: an updated review, Energy Sustain Dev 14 (2010) 302–314.

[26] R. Doherty, M. O'Malley, A new approach to quantify reserve demand in systems with significant installed wind capacity, IEEE Trans Power Syst 20 (2005) 587–595.

[27] H. Chen, T.N. Cong, W. Yang, C. Tan, Y. Li, Y. Ding, Progress in electrical energy storage system: a critical review, Prog Nat Sci 19 (2009) 291–312.

[28] S.H. Jensen, C. Graves, M. Mogensen, C. Wendel, R. Braun, G. Hughes, et al. Large-scale electricity storage utilizing reversible solid oxide cells combined with underground storage of CO_2 and CH_4, Energ Environ Sci 8 (2015) 2471–2479.

[29] S. Becker, B.A. Frew, G.B. Andresen, T. Zeyer, S. Schramm, M. Greiner, et al. Features of a fully renewable US electricity system: optimized mixes of wind and solar PV and transmission grid extensions, Energy 72 (2014) 443–458.

[30] H. Zhao, Q. Wu, S. Hu, H. Xu, C.N. Rasmussen, Review of energy storage system for wind power integration support, Appl Energ 137 (2015) 545–553.

[31] P.D. Brown, J.A. Peas Lopes, M.A. Matos, Optimization of pumped storage capacity in an isolated power system with large renewable penetration, IEEE Trans Power Syst 23 (2008) 523–531.

[32] M. Swierczynski, R. Teodorescu, C.N. Rasmussen, P. Rodriguez, H. Vikelgaard, Overview of the energy storage systems for wind power integration enhancement, IEEE (2010) 3749–3756.

[33] A. Choudhury, H. Chandra, A. Arora, Application of solid oxide fuel cell technology for power generation—a review, Renew Sustain Energy Rev 20 (2013) 430–442.

[34] Q. Fu, C. Mabilat, M. Zahid, A. Brisse, L. Gautier, Syngas production via high-temperature steam/CO_2 co-electrolysis: an economic assessment, Energy Environ Sci 3 (2010) 1382–1397.

[35] Y. Luo, X. Wu, Y. Shi, A.F. Ghoniem, N. Cai, Exergy analysis of an integrated solid oxide electrolysis cell-methanation reactor for renewable energy storage, Appl Energ 215 (2018) 371–383.

[36] Y. Luo, Y. Shi, N. Cai, Power-to-gas energy storage by reversible solid oxide cell for distributed renewable power systems, J Energ Eng 144 (2018) 4017079.

[37] T. Ma, H. Yang, L. Lu, Development of hybrid battery–supercapacitor energy storage for remote area renewable energy systems, Appl Energ 153 (2015) 56–62.

[38] J. Zhu, W. Yuan, M. Qiu, B. Wei, H. Zhang, P. Chen, et al. Experimental demonstration and application planning of high temperature superconducting energy storage system for renewable power grids, Appl Energ 137 (2015) 692–698.

[39] M. Justo Alonso, P. Liu, H.M. Mathisen, G. Ge, C. Simonson, Review of heat/energy recovery exchangers for use in ZEBs in cold climate countries, Build Environ 84 (2015) 228–237.

[40] L. Zhang, W.L. Lee, Evaluating the use heat pipe for dedicated ventilation of office buildings in Hong Kong, Energ Convers Manage 52 (2011) 1983–1989.

[41] H. Zeng, Y. Wang, Y. Shi, N. Cai, D. Yuan, Highly thermal integrated heat pipe-solid oxide fuel cell, Appl Energ 216 (2018) 613–619.

[42] H. Zeng, Y. Wang, Y. Shi, N. Cai, X. Ye, S. Wang, Highly thermal-integrated flame fuel cell module with high temperature heatpipe, J Electrochem Soc 164 (2017) F1478–F1482.

[43] W. Pu, C. Yue, D. Han, W. He, X. Liu, Q. Zhang, et al. Experimental study on Organic Rankine cycle for low grade thermal energy recovery, Appl Therm Eng 94 (2016) 221–227.

[44] H. Nami, I.S. Ertesvåg, R. Agromayor, L. Riboldi, L.O. Nord, Gas turbine exhaust gas heat recovery by organic Rankine cycles (ORC) for offshore combined heat and power applications—energy and exergy analysis, Energy 165 (2018) 1060–1071.

[45] S. Lecompte, H. Huisseune, M. van den Broek, B. Vanslambrouck, M. De Paepe, Review of organic Rankine cycle (ORC) architectures for waste heat recovery, Renew Sustain Energy Rev 47 (2015) 448–461.

[46] J. Park, H. Lee, M. Bond, Uninterrupted thermoelectric energy harvesting using temperature-sensor-based maximum power point tracking system, Energ Convers Manage 86 (2014) 233–240.

[47] X. Niu, J. Yu, S. Wang, Experimental study on low-temperature waste heat thermoelectric generator, J Power Sources 188 (2009) 621–626.

[48] X. Liu, Y.D. Deng, Z. Li, C.Q. Su, Performance analysis of a waste heat recovery thermoelectric generation system for automotive application, Energ Convers Manage 90 (2015) 121–127.

[49] F.J. Lesage, N. Pagé-Potvin, Experimental analysis of peak power output of a thermoelectric liquid-to-liquid generator under an increasing electrical load resistance, Energ Convers Manage 66 (2013) 98–105.

[50] C. Babu, P. Ponnambalam, The role of thermoelectric generators in the hybrid PV/T systems: a review, Energ Convers Manage 151 (2017) 368–385.

[51] J. Song, Y. Song, C. Gu, Thermodynamic analysis and performance optimization of an Organic Rankine Cycle (ORC) waste heat recovery system for marine diesel engines, Energy 82 (2015) 976–985.

[52] B. Orr, A. Akbarzadeh, M. Mochizuki, R. Singh, A review of car waste heat recovery systems utilising thermoelectric generators and heat pipes, Appl Therm Eng 101 (2016) 490–495.

[53] M.Q. Raza, A. Khosravi, A review on artificial intelligence based load demand forecasting techniques for smart grid and buildings, Renew Sustain Energy Rev 50 (2015) 1352–1372.

[54] C. Deb, F. Zhang, J. Yang, S.E. Lee, K.W. Shah, A review on time series forecasting techniques for building energy consumption, Renew Sustain Energy Rev 74 (2017) 902–924.

[55] F. Shariatzadeh, P. Mandal, A.K. Srivastava, Demand response for sustainable energy systems: a review, application and implementation strategy, Renew Sustain Energy Rev 45 (2015) 343–350.

[56] L.Y. Seng, G. Lalchand, G.M. Sow Lin, Economical, environmental and technical analysis of building integrated photovoltaic systems in Malaysia, Energ Policy 36 (2008) 2130–2142.

[57] M. Yoda, S. Inoue, Y. Takuwa, K. Yasuhara, M. Suzuki, Development and commercialization of new residential SOFC CHP system, ECS Trans 78 (2017) 125–132.

[58] Maruta A. Japan's ENE-FARM programme; 2016. https://www.energyagency.at/fileadmin/dam/pdf/veranstaltungen/Brennstoffzellenworkshop_Oktober/Maruta.pdf.

[59] E. Unamuno, J.A. Barrena, Hybrid ac/dc microgrids—Part II: Review and classification of control strategies, Renew Sustain Energy Rev 52 (2015) 1123–1134.

[60] J. Cao, K. Meng, J. Wang, M. Yang, Z. Chen, W. Li, et al. An energy internet and energy routers, Sci China Inform Sci 44 (2014) 714–727.

[61] Z. Dong, J. Zhao, F. Wen, Y. Xue, From smart grid to energy internet: basic concept and research, Autom Electric Power Syst 38 (2014) 1–11.

Chapter 3

Bridging a bi-directional connection between electricity and fuels in hybrid multienergy systems

Yu Luo[a], Yixiang Shi[b] and Ningsheng Cai[b]

[a]National Engineering Research Center of Chemical Fertilizer Catalyst (NERC-CFC), Fuzhou University, Fujian, China; [b]Department of Energy and Power Engineering, Tsinghua University, Beijing, China

3.1 Introduction

Distinguish with centralized energy systems, distributed hybrid systems (DHSs) offer a user-oriented multienergy supply pattern with flexible scale and have the ability to meet diverse consumer demands. Therefore, we believe that DHS is the basic unit of distributed energy networks and the forthcoming Energy Internet. Increasing energy demands and diverse energy utilization preferences require DHS to develop with the following trends:

1. *Going Green* [1]: A number of countries have made policy and regulatory changes, aiming to increase green energy and reduce greenhouse gas emissions.
2. *Multienergy comprehensive management and efficiency improvement*: Multienergy resources and demands require DHS to dispatch multienergy flow more remarkably and realize energy cascade utilization to reduce energy losses and achieve higher efficiency.
3. *Faster response*: The rise of intermittent renewable energy sources and the diversity of end-use consumers need DHS to response more quickly to the real-time variation of distributed energy sources and energy demands.
4. *Compacter*: DHS at a smaller scale needs a compactor system integration to make it available for households or specific buildings.
5. *Peer-to-peer energy transactions* [1]: More and more end-use energy consumers are becoming the "prosumers," who is now also energy producers.

Hybrid Systems and Multi-energy Networks for the Future Energy Internet.
http://dx.doi.org/10.1016/B978-0-12-819184-2.00003-1

As mentioned in Chapter 2, an advanced DHS could include energy generation subsystem, energy storage subsystem, energy recovery subsystem and energy end-use subsystem. To follow these development trends, DHS should integrate more promising advanced energy conversion devices into these four subsystems, such as advanced fuel cells and advanced energy storage technologies, forming ordered and cascaded energy flows. In particular, a flexible conversion between electricity and fuels is significant technical support for the decentralization of the energy systems. On one hand, fuels can be stored seasonally without any self-charging losses. On the other hand, fuels, particularly liquid fuels, can be transported through well-developed transportation network to form connections between DHS. To realize a flexible conversion between electricity and fuels, advanced fuel cells and electrolyzers are the key technologies.

In this chapter, we introduce the working principles of fuel cells and electrolyzers, as well as their contributions to DHS. For each technology, current status, potential applications, and challenges are comprehensively reviewed and discussed.

3.2 Fuel cells for energy generation

Conventional power generation based on heat engines first burns fuels into high-grade thermal energy, then converting thermal energy into kinetic energy, and further driving electric generator to generate electric energy. Heat engines rely on pressure and temperature gradients are limited by the Carnot efficiency. A fuel cell is an electrochemical device that is capable of converting chemical energy directly into electrical energy without the limitation of Carnot efficiency. Anode, electrolyte, and cathode are three essential component of a fuel cell, as Fig. 3.1A shows. Oxidation half reaction happens in anode, while reduction half reaction happens in cathode. Fuel losses electrons in anode, then the released electrons transport to cathode through external circuit and combine with oxidant (typically air or O_2).

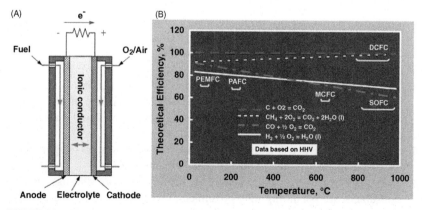

FIGURE 3.1 Working principle (A) and major types (B) of fuel cell technologies. *(Modified from Ref. [2] Copyright 2012 Elsevier Ltd.)*

Meanwhile, electrolyte transferred ions released or consumed in both electrodes to form a continuous current flow. In comparison with batteries, fuel cells can continuously supply power only if enough fuel supply. Because of no Carnot efficiency limitation, fuel cells can still remain operating with a high efficiency even at a small scale, which offers a distinct advantage over conventional heat engines. Particularly, a fuel cell with balance-of-plant (BOP) can reach an overall efficiency of ~90% when applied for combined heat and power (CHP) or combined cold, heat, and power (CCHP) systems. Besides, direct fuel-to-electricity lowers the emissions of pollutant (SO_2, NOx, particulate matters, etc.) and CO_2, reduce infrastructure space and operating noisy, as well as has the ability of fast start-up and response. As a consequence, fuel cells can meet many aspects of requirement for DHS, showing great application potential in DHS.

3.2.1   Fuel cell efficiency and classification

In thermodynamics, the maximum electricity generation in a fuel cell is generally equal to the negative Gibbs free energy $-\Delta G$ of the overall reaction, that is, fuel oxidation reaction. The chemical energy containing in fuel is generally represented by heat value (HV), which is the negative enthalpy change $-\Delta H$ of the fuel oxidation reaction. Consequently, the theoretical efficiency of the electrochemical oxidation in a fuel cell can be represented as follows:

$$\eta_{th} = \frac{\Delta G}{\Delta H} \tag{1}$$

The theoretical efficiency η_{th} denotes the maximum fuel-to-electricity efficiency in a fuel cell. The Gibbs free energy ΔG and the enthalpy change ΔH have the following relation:

$$\Delta G = \Delta H - T\Delta S \tag{2}$$

where ΔS is the entropy change of the fuel oxidation reaction and T is the temperature. Generally, the value of ΔS for a fuel oxidation reaction is negative. Thus, thermodynamics reveals that a part of chemical energy in fuel inevitably converts to thermal energy in a fuel cell. The minimum value of heat generation is equal to $T\Delta S$, which is called as Reversible Heat. Fig. 3.1B shows the variation of η_{th} with operating temperature for fuel cells fueled with various fuels. The white solid line refers to η_{th} of an H_2-fueled fuel cell, which indicates that η_{th} decreases with operating temperature. The theoretical efficiency η_{th} is equal to 83% at room temperature (25°C), while η_{th} drops to 76% at 800°C.

 Apparently, a real electrochemical system hardly achieves a theoretical efficiency due to inevitable polarization losses. According to the maximum electricity generation (represented by $-\Delta G$), the equilibrium potential E_N, that is, the theoretical open circuit voltage (OCV), can be expressed as follows:

$$E_N = -\frac{\Delta G}{nF} \tag{3}$$

where n is the number of electrons transferred in the overall fuel oxidation reaction, and F is the Faraday constant (96,485 C mol^{-1}). An actual fuel cell has to operate at a voltage lower than E_N to overcome irreversible losses and offer a higher current density. The irreversible losses typically include activation polarization losses, ohmic polarization losses, and concentration polarization losses. The operating voltage V_{cell} can be usually expressed as:

$$V_{cell} = E_N - V_{act} - V_{ohm} - V_{conc} - V_{leak} \quad (4)$$

Here, V_{act}, V_{ohm}, V_{conc}, and V_{leak} are the activation polarization voltage, ohmic polarization voltage, concentration polarization voltage, and leak voltage (V), respectively. The activation polarization voltage V_{act} is used to activate electrochemical reactions and accelerate electrons generation (in anode) and consumption (in cathode), which is generally remarkable at low current density. The ohmic polarization voltage V_{ohm} is to overcome the electronic resistance in electrode, interconnects or current collectors and the ionic resistance in electrolyte, which linearly increases with current density. The concentration polarization voltage V_{conc} is to overcome the species difference between bulk concentration and concentration at electrochemical reaction interface caused by mass diffusion resistance. The leak voltage V_{leak} is voltage deviation resulting from some extra issues like gas leakage, leaky electrolyte, etc. Considering a series of irreversible losses, the voltage efficiency η_V is defined as below:

$$\eta_V = \frac{V_{cell}}{E_N} \quad (5)$$

Based on Eqs. (1), (3), and (5), the overall electrical efficiency of a fuel cell η_{el} can be obtained according to the following expression:

$$\eta_{el} = \eta_V \eta_{th} = \frac{nFV_{cell}}{\Delta H} \quad (6)$$

Here, ΔH generally varies little with temperature, therefore, η_{el} is strongly relied on operating voltage. However, in actual operation, the selection of operating voltage greatly depends on the electrochemical performance of a fuel cell, which has a close correlation with operating temperature.

Fig. 3.2 shows the typical polarization curve, power density curve, and corresponding efficiencies. The polarization curve shows a sharper drop in voltage at low current density or high current density, which is caused by activation polarization and concentration polarization, respectively. The fuel cell obtains the maximum power density at ~0.45 V. According to the polarization curve, the theoretical efficiency η_{th} is 86% assuming no leakage voltage. The voltage efficiency η_V is equal to 100% at OCV, and drops with decreasing operating voltage (rising polarization voltage). The overall electrical efficiency η_{el}, that is,

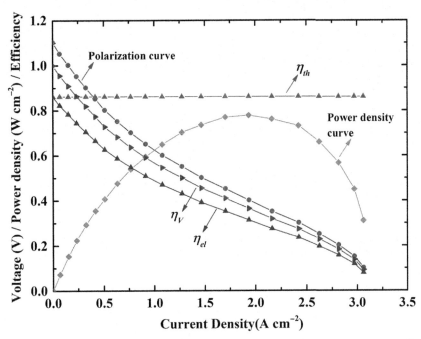

FIGURE 3.2 Typical polarization curve, power density curve, and corresponding efficiencies.

the product of η_{th} and η_V, is 86% at OCV, and drops to 35% at 0.45 V (the peak power density). Generally, fuel cells can operate at ~0.7 V for a long time, that is, operating at an overall electrical efficiency of 55%. The development in advanced materials for fuel cells allows to increase operating voltage and further enhance the overall electrical efficiency.

Up to now, there are still not any types of fuel cells capable of operating in a wide temperature range from room temperature or lower to ~1000°C. According to different operating temperature range, Fig. 3.1B marks the theoretical efficiency range of various fuel cells. As marks in Fig. 3.1B, various types of fuel cells are under development and potentially applicable at different temperature. The general classification of fuel cells is based on the electrolyte materials. Current fuel cell technologies under development mainly including proton exchange membrane fuel cell (PEMFC), alkaline fuel cell (AFC), phosphoric acid fuel cell (PAFC), molten carbonate fuel cell (MCFC), solid oxide fuel cell (SOFC), etc. However, Fig. 3.1B shows fuel cell technologies applicable in the temperature range of 300–600°C are still lacking. To fill up this gap, researchers have paid their efforts on extending the temperature range of these fuel cell technologies. For example, novel polymer membranes are potential to extend the operating temperature of PEMFC above 100°C (up to ~200°C) [3]. AFC using molten alkaline hydroxide electrolyte is proven to enhance the operating temperature to higher than 450°C [4,5]. Proton-conducting ceramic electrolytes are able to lower the operating temperature of SOFC to 350°C [6].

Table 3.1 shows the half reactions, electrolyte, and suitable operating temperature of various fuel cell technologies. According to the typical operating temperature, different fuel cells can be classified into low temperature fuel cells (<100°C), medium temperature fuel cells (200–600°C), and high temperature fuel cells (>600°C). Typically, conventional LT-PEMFC and aqueous or anion exchange membrane (AEM) AFC belong to low temperature fuel cells. HT-PEMFC, molten hydroxide AFC, PAFC, and proton-conducting SOFC (H-SOFC) belong to medium temperature fuel cells. MCFC and conventional oxygen-conducting SOFC (O-SOFC) are high temperature fuel cells. Differences in reaction mechanism, materials, and operating temperature among various types of fuel cells lead to diverse characteristics, application scenes, advantages, and unsolved issues, as summarized in Table 3.2. Here, we only discuss three typical fuel cells: PEMFC, AFC, and SOFC.

TABLE 3.1 Typical fuel cells under development (classified by electrolyte types) [2].

Fuel cell type	Fuel anode reaction	Electrolyte	Cathode reaction	Temp. (°C)
PEMFC	H_2 $H_2 \rightarrow 2H^+ + 2e^-$	PEM $H^+ \rightarrow$	$1/2O_2 + 2H^+ +$ $2e^- \rightarrow H_2O$	60–100 (LT-PEMFC) 100–200 (HT-PEMFC)
AFC	H_2 $H_2 + 2OH^- \rightarrow$ $2H_2O + 2e^-$	$Me(OH)_x$ $OH^- \leftarrow$	$1/2O_2 + H_2O +$ $2e^- \rightarrow 2OH^-$	<100 (Aqueous/AEM) 200–550 (Molten hydroxide)
PAFC	H_2 $H_2 \rightarrow 2H^+ + 2e^-$	H_3PO_4 $H^+ \rightarrow$	$1/2O_2 + 2H^+ +$ $2e^- \rightarrow H_2O$	160–220
MCFC	CH_x, CO, H_2 $H_2 + CO_3^{2-} \rightarrow H_2O +$ $CO_2 + 2e^-$	Molten Me_xCO_3 $CO_3^{2-} \leftarrow$	$1/2O_2 + CO_2 +$ $2e^- \rightarrow CO_3^{2-}$	600–800
SOFC	CH_x, CO, H_2 $H_2 + O^{2-} \rightarrow H_2O$ $+ 2e^-/H_2 \rightarrow 2H^+$ $+ 2e^-$	Ceramic $O^{2-} \leftarrow /$ $H^+ \rightarrow$	$1/2O_2 + 2e^- \rightarrow$ $O^{2-}/1/2O_2$ $+ 2H^+ +$ $2e^- \rightarrow H_2O$	600–1000 (O-SOFC) 400–800 (H-SOFC)

TABLE 3.2 Comparison in general characteris among various fuel cells [2,7].

Fuel cell type	PEMFC	AFC	PAFC	MCFC	SOFC
Typical stack size (kW)	<1–100	1–100	5–400	300–3000	1–2000
Electrical efficiency (LHV)	60% (direct H_2) 40% (reformed fuel)	45%–60%	40%–45%	45%–55%	50%–65%
Thermal insulation	Low	Low	Medium	High	High
Internal reforming	Not possible	Not possible	Not possible	Only with steam	Only with steam
Impurity sensitivity	CO/S ~20 ppm NH_3 ~0.1 ppm	CO_2/CO/S	Sulfur	Sulfur	Sulfur
Power at cold start	>50%	>50%	—	—	—
Thermal cycling	Unlimited	Unlimited	Good	Restricted	Restricted
Start-up/shut-down	Very fast (s)	Very fast (s)	Slow (h)	Several hours	<1 h to several hours
Load following	Excellent	Excellent	Limited	Limited	Limited
BOP	Simple	Simple	Medium	Complex	Complex
Application	Backup power; portable power; distributed generation; transportation; specialty vehicles	Military; space; backup power; transportation	Distributed generation	Electric utility; distributed generation	Auxiliary power; electric utility; distributed generation
Advantages	Low temperature; alleviated corrosion and electrolyte management problems; quick start-up and load following	Lower cost components; low temperature; quick start-up	Suitable for CHP; increased tolerance to fuel impurities	High efficiency; fuel flexibility; suitable for CHP; hybrid/gas turbine cycle	High efficiency; fuel flexibility; solid electrolyte; suitable for CHP; hybrid/gas turbine cycle
Challenges	Expensive catalysts; sensitive to fuel impurities	Sensitive to CO_2 in fuel and air; electrolyte management (aqueous); electrolyte conductivity (AEM)	Expensive catalysts; long start-up time; sulfur sensitivity	High temperature corrosion and breakdown of cell components; long start-up time; low power density	Long start-up time; limited number of shutdowns; mechanical failure

3.2.2 Proton exchange membrane fuel cell

PEMFC uses polymer electrolyte for conducting protons. Typical PEMFC electrolyte is perfluorosulfonic acid (PFSA) polymer, for example, Nafion®, which exhibits high proton conductivity, high chemical stability, good mechanical strength, and high flexibility [3]. Currently, PFSA-based PEMFC is the most widely used fuel cell technology. Anode and cathode are paint on both sides of PFSA polymer membrane electrolyte into a membrane electrode assembly (MEA), which is known as the key component of commercially applicable PEMFCs. To scale up PEMFCs, electronic-conducting interconnect plates connect all the MEAs in series to form an PEMFC stack. The electrochemical performance of PEMFC can be greatly affected by the temperature and humidity of PFSA polymer. To obtain better performance, PFSA-based PEMFC should operate at a relatively low operating temperature (up to 80°C) at atmospheric pressure for maintaining the membrane humid. A dry PFSA polymer membrane can leads to the proton conductivity significantly dropping, while excess water may reduce electrochemical reaction kinetics due to the electrode/electrolyte interface flooding. Therefore, PEMFC stacks need heat and water management systems.

The electrical efficiency of PEMFC is 40%–60%. When high-purity H_2 is fueled, PEMFC can operate at a higher efficiency ($\sim$60%). For other hydrogen-rich fuels such as methane, methanol, gasoline, ammonia, and so on, it is not possible to integrate internal reforming into PEMFC stack because of low operating temperature, hence, an extra fuel processing system is needed to produce PEMFC-tolerable hydrogen-rich fuel gas. The relative low hydrogen content reduces the electrical efficiency to $\sim$40%. Nevertheless, the system efficiency (including electrical energy and thermal energy) can reach 60%–80% if low-grade heat can be effectively recovered by energy recovery subsystem. Low operating temperature of PEMFC contributes to maintain over 50% of rated power even at cold start as well as have unlimited thermal cycling, fast start-up or shut-down, excellent load following, and simple BOP.

However, low operating temperature causes PEMFC suffering CO-poisoning even at ppm-level CO partial pressure. To alleviate this issue, high temperature PEMFC, typically based on acid-based polymer membranes, is proposed and under development to elevate the operating temperature to 200°C [3]. Besides, slow oxygen reaction kinetics, expensive catalysts (due to high noble metal loading) and membrane degradation are also the challenges PEMFC is facing with.

Because of these features, the application of PEMFC is quite flexible, involving backup/un-interruptible power, portable power, distributed generation (such as residential micro-CHP and remote area power supplies), transportation (including cars, buses, trucks, locomotives, submarines, wheel chairs, auto-bicycles, delivery vans, armored vehicles, and small transporters at airports, shipyards, and railway stations), specialty vehicles, etc. [2,7]. Typical size for an PEMFC stack varies from less than 1 to ~100 kW [7]. For stationary application, current commercial PEMFCs generally have capacities of up to 10 kW. Osaka Gas and Tokyo gas began to sell the PEMFC-type system "ENE-FARM" since 2009 [8], and has successfully sold over 140,000 PEMFC units from 2009 to 2015. Their 3rd generation PEMFC-based residential CHP systems can offer a power generation output of 200–750 W with electrical efficiency of 39% (LHV) and overall system efficiency of 95% [9]. The price of PEMFC-based units has reduced by two thirds to ~JPY 1 million, and is expected to drop to JPY 800,000 by 2019 [10]. The portable, mobile, or automotive application usually requires fuel cells with better behaviors in start-up, system compactness (volume and weight), mechanical robustness, load following, etc. PEMFC is suitable for these applications. In electric vehicles, 100 kW-level PEMFCs have been commercially available. The new Mirai fuel cell vehicle uses a single line PEMFC stack with 370 cells to offer a maximum power output of 114 kW and achieve a volumetric power density of 3.1 kW L^{-1} and a mass power density of 2.0 kW kg^{-1} [11]. By the end of 2017, Toyota announced that the annual sales of Mirai fuel cell reached 2700 units and the total sales reached 5300 units.

3.2.3 Alkaline fuel cell

AFC, that is, Bacon fuel cell, is one of the earliest and most developed fuel cell technologies. Since 1960s, NASA has used AFC for Apollo-series space missions for producing potable water, low-grade heat, and electricity. Conventional AFC uses aqueous KOH/NaOH solution with mass fraction of 35–85 wt.% as the liquid electrolyte, and generally operates at less than 100°C to avoid water loss from the alkaline electrolyte solution. Therefore, AFC has some common advantages with PEMFC, including low thermal insulation requirement, unlimited thermal cycling, second-level start-up or shut-down, excellent load following, simple BOP, cold start, etc. Moreover, AFC only needs to use non-noble metal catalysts (typically nickel) as electrode materials, thereby reducing component cost remarkably. Similarly to PEMFC, AFC can be poisoned by CO and S. Furthermore, alkaline electrolyte is quite sensitive to CO_2 existing in fuel or oxidant stream, leading to AFC suffering a dramatic degradation due to hydroxide ions decreasing and porous electrode blocking (metal carbonate precipitating). Removing CO_2 from inlet fuel and oxidant stream or recycling electrolyte for outside scrubbing is alternative solution to address this issue. Nevertheless, these solutions seem hard to solve the degradation issue absolutely when air from atmosphere is fed into AFC as the oxidant [12]. Besides, AFC needs extra electrolyte management for avoiding the corrosion of aqueous alkaline electrolyte and electrode flooding or drying.

To solve the issues related to the aqueous alkaline electrolyte, AFC using solid AEM attracts rising attentions. AEM-based AFC not only avoids liquid electrolyte management system, but also has faster electrochemical kinetics of the oxygen reduction reaction and wider selections for valuable chemicals as by-products [13]. In addition, liquid fuels like methanol, ethanol, formate, and liquid ammonia are capable of driving AEM-based AFC for electricity generation directly. Nevertheless, conducting OH^- via membrane is much slower than conducting H^+ due to large molecule weight. To enhance the conductivity of OH^-, polymer in AEM needs to increase, leading to decreasing mechanical strength [13]. In addition, the stability of AEM is still a challenge particularly in CO_2-containing atmosphere and elevated temperature.

Aqueous AFC is the cheapest of all fuel cell technology to manufacture. The major markets for AFC include military, space shuttles, backup power, and transportation. Typical AFC stack size varies in the range from ~ 1 to ~ 100 kW. Siemens and UTC successfully operated their respective AFC stack by feeding pure hydrogen and oxygen (CO_2-absence) for 20,000 and 15,000 h, respectively [14]. AFC Energy from UK is the leading AFC power company, who have built the world's largest operational AFC systems called KORE in Stade, Northern Germany to offer over 200 kW electricity [15]. KORE system consists of three tiers and each tier includes eight AFC stacks. Each AFC stack in KORE system has a rated power of 10 kW with 101 cells. In January 2019, AFC Energy announced that they have powered up world's first hydrogen electric vehicle charger called CH_2ARGE^{TM} based on their AFC technology to offer 100% clean recharging for electric vehicles [16]. AFC Energy announced that their next generation AFC stack will use AEM as the electrolyte and also have a design power output of 10 kW [17]. Another project carried out by AFC Energy is AL-KAMMONIA, aiming to develop a proof-of-concept ammonia-fueled system for power supply in remote area [18] by integrating their AFC stacks, efficient NH_3 decomposition system and novel ammonia fuel system. Unlike PEMFC, AFC has high NH_3 tolerance due to its alkaline electrolyte, even has potential to convert NH_3 directly to electricity, that is, direct ammonia fuel cell (DAFC). DAFC will be specifically discussed in Section 3.

3.2.4 Solid oxide fuel cell

SOFC is an all-solid-state fuel cell based on ceramic electrolyte. Up to now, commercially available SOFCs use oxygen-ion conducting ceramic electrolyte, commonly yttria stabilized zirconia (YSZ). YSZ-based SOFC prefers to operate at 600–1000°C. High operating temperature enables to use nonnoble metal as the electro-catalyst. Generally, SOFC anode uses Ni-YSZ cermet, where Ni is both the electronic conductor and electro-catalyst and YSZ is the ionic conductor. SOFC cathode uses a ceramic composite mixing YSZ with doped lanthanum manganite (LSM) [19]. Differently from most membrane-based fuel cells, the mechanical strength of SOFC typically relies on the Ni-YSZ anode with a thickness of hundreds of micrometers and the electrolyte only tens of micrometers to lower ohmic resistance. High operating temperature not only allows the use of nonnoble metal catalysts, but also enhances the electrical efficiency and impurity tolerance of SOFC. SOFC operates at an electrical efficiency of

50%–65%, in the meantime, high-grade heat is generated. In a micro-CHP system integrating SOFCs, the overall system efficiency can exceed 90%. SOFCs don't suffer CO-poisoning, moreover, CO is an important fuel for SOFC. Therefore, internal reforming is possible in SOFCs, enlarging the fuel flexibility of SOFC. However, hydrocarbon fuels may result in severe carbon deposition, leading to catalyst deactivation, pore blocking, and mechanical failure [20]. Adding reforming catalyst layer on the anode surface and developing novel anticarbon anode materials are both alternative solutions to avoid carbon deposition [6,21–24]. In addition, sulfur containing in fuel stream is the main impurity poisoning SOFC. Only several ppm of H_2S can lead to remarkable degradation of SOFCs [25]. Consequently, design for sulfur-tolerant anode materials is one of the important directions for SOFC research.

Common configurations of an SOFC unit are planar SOFC and tubular SOFC, as Fig. 3.3 depicts. Planar SOFCs are flat and separated by electron-conducting interconnectors to connect with each other in series. Fuel stream and oxidant stream flow in two sides of the planar SOFC, respectively. The structure of a planar SOFC stack is similar to an PEMFC stack. Tubular SOFCs are generally through-hole or dead-end tubes with inside anode and outside cathode. Fuel stream flows inside the tube and oxidant stream flows outside the tube. Planar SOFCs have the advantages of high current density and easy manufacture, but suffer a series of issues including difficult sealing, low mechanical strength and low thermal cycle resistance. On the contrary, tubular SOFCs have relatively low power density and complex manufacturing processes, while have higher mechanical strength, higher thermal cycle resistance, and faster start-up than planar SOFCs. Moreover, tubular SOFCs are easy for sealing and stack assembly. Even sealing is not required when the outlet gas from both electrodes are directly mixed and burned. Slow start-up (hour-level) and limited load following are considered to be the shortages of SOFC due to its high operating temperature. Too fast temperature variation could lead to the mechanical failure of SOFCs. Therefore, the applications of SOFC in electric vehicle and portable power are limited. Micro-tubular SOFCs proposed by K. Kendall show high thermal shock resistance and their sufferable temperature variation is up to 200–550°C min^{-1} [28], making SOFC comparable with low-temperature or medium-temperature fuel cells.

Conventional SOFCs mainly aim at stationary power generation with the sizes varying from ~1 kW to 2 MW. Common application scenes of SOFCs include auxiliary power, electric utility, and distributed generation (typically residential or industrial CHP or CCHP). SOFC can be integrated with microgas turbines or low pressure stream turbines to utilize high-grade waste heat. Westinghouse and later Siemens-Westinghouse began the early demonstration of SOFC system in the late 1980s and early 1990s [2]. Up to now, scaled demonstration SOFC plants with size of over 50 kW have successively constructed or carried out by Siemens-Westinghouse, Mitsubishi Heavy Industries, Rolls Royce Fuel Cells, and Bloom Energy, while small-scale SOFC-based CHP units are demonstrated by Sulzer-Hexis, Ceres Power, Ceramic Fuel Cells Limited, Topso Fuel Cells, Delphi,

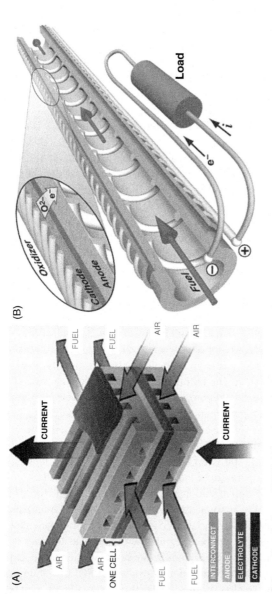

FIGURE 3.3 Schematics of planar SOFC (A) and tubular SOFC (B). *(Modified from Ref. [26] Copyright 2013 Elsevier Ltd. and Ref. [27] Copyright 2007 Elsevier Ltd.)*

Acumentrics, Fuel Cell Technologies, Global Thermoelectric Inc., ZTEK Corp, etc. [2]. Arguably, ENE-FARM program in Japan could be the most successful fuel cell commercialization program for residential application using PEMFC and SOFC. ENE-FARM program aims to develop residential co-generation systems with capacities of 0.2–3 kW, carried out by a consortium involving Kyocera Corporation, Tokyo Gas, Osaka Gas, Aisin Seiki Co. Ltd., Chofu Seisakusho Co. Ltd., Toyota Motor Corporation, etc. [29]. Despite slower commercialization and higher price, SOFC-based ENE-FARM units show higher efficiency and compatibility with residential application than PEMFC. Two types of natural gas-fueled SOFC units, that is, 700 W-type and 3 kW-type, have been commercialized in Japan with a rated electrical efficiency of 52% and overall system efficiency of >90% [30,31]. At present, over 220,000 ENE-FARM units have been installed since 2009 and the percentage of SOFC-based units in total installed units also rises from 2% to >10% over the period from 2011 to 2017 [10,31]. Nissan Motor Co., Ltd. developed the world's first SOFC-powered vehicle system fueled with bio-ethanol [32]. Unlike PEMFC vehicle system, Nissan's SOFC vehicle system uses battery to drive the motor and SOFC to charge the battery, hence, the hybrid SOFC-battery system can achieve high-efficiency and fast-response operation.

3.2.5 Fuel cells fueled with diverse fuels

Hydrogen is an ideal fuel, showing wide compatibility with most kinds of fuel cells mentioned above expect direct alcohol fuel cells (requiring either ethanol or methanol), direct borohydride fuel cells [2]. However, H_2 has a low boiling point ($-253°C$) and a low volumetric energy density (0.0108 MJ L^{-1}) at 25°C and 1 atm. H_2 storage is quite difficult because of the high diffusibility, buoyancy, and activity of H_2. A common approach is to pressurize H_2 to 700 bar, increasing the volumetric energy density to ~6 MJ L^{-1} [33]. But this approach could cause an increase of risk level in terms of H_2 storage. Another solution is to use an alternative fuel for fuel cells. A system integrating fuel processing subsystem and H_2-fueled PEMFC subsystem can convert energy-dense liquid fuel indirectly to electricity, which is closer to commercialization and more economical for the kW-scale or larger application due to developed fuel reforming and PEMFC technologies. However, extra processes are needed to remove CO, sulfur, or NH_3 to avoid PEMFC poisoning. IdaTech from US has used reformed methanol fuel cell as a backup power for telecommunication base stations. Their "extended run'" system fueled with 59 US gallon methanol-water mixture (equivalent to 25 standard H_2 cylinders) can continuously operate for 40 h at a power output of 5 kW [34]. Small-scale demonstrations of ammonia-fueled fuel cell systems are also performed. STFC Rutherford Appleton Laboratory cooperating with University of Oxford integrates ammonia catalytic cracking reactor (using lithium imide catalyst) with NH_3 removal process (using anhydrous $MgCl_2$ power) and 100 W-level PEMFC to power a TV [35]. A more compact system was demonstrated by Korea Institute of Science and Technology (KIST).

KIST integrates NH_3 catalytic decomposition (using $Ru/La-Al_2O_3$ catalyst) and NH_3 removal (using 13X zeolite) with 1 kW-level PEMFC and BOP like heat exchangers to power a wire-linked unmanned aerial vehicle (UAV) [36]. This system uses a 3.4 L NH_3 tank and is capable of improving the continuous operating time of UAV from 14 min to ∼4.1 h. As KIST expected, the gravimetric energy density significantly increases with system weight (<50 kg) owing to the mass fraction of ammonia tank in the whole system rising. The ammonia-fueled fuel cell system exceeds lithium-ion batteries in energy density when the system is over 21 kg in weight [36].

Apart from indirect fuel utilization, a wide range of fuels have shown their potential compatibilities with fuel cells, such as carbon monoxide or syngas, natural gas, methanol, ethanol, gasoline, diesel, coal, and ammonia. Nevertheless, there are limited fuels directly available for commercialized fuel cells except H_2, CO, or syngas [2]. Therefore, these fuels can be converted into H_2 or syngas then indirectly fueled for fuel cells via extra fuel processing systems such as catalytic reforming, catalytic partial oxidation, or catalytic decomposition. As Table 3.3 reveals, most fuel cells are sensitive to S-containing impurities more or less. A desulfurizer is generally required when fuel cells are fueled by S-containing fuels. In addition, low temperature or medium temperature fuel

TABLE 3.3 Thermodynamic features (at 25°C) of various fuels applicable or potentially applicable for direct integration with fuel cells [38,39].

Fuels	M_w (g mol^{-1})	n	n/M_w (mol e$^-$ g^{-1})	E^0 (V)	Energy density		η_{th} (%)
					(Wh kg^{-1})	(Wh L^{-1})	
Hydrogen (700 bar)	2.01	2	0.995	1.23	32,802	1378	83
Methane (250 bar)	16.04	8	0.499	1.04	13,889	2490	100
Methanol	32.04	6	0.187	1.21	6073	4820	97
Ethanol	46.07	12	0.260	1.20	8377	6280	95
Formic acid	46.03	2	0.043	1.40	1630	1750	106
Dimethyl ether	46.07	12	0.260	1.20	8377	5610	95
Ammonia (>8 bar)	17.03	3	0.176	1.17	5529	3411	103
Hydrazine	32.05	4	0.125	1.62	5419	5400	100
Ammonia borane	30.87	6	0.194	1.62	8400	6100	–
Coal/carbon	12.01	4	0.333	1.02	9133	5800–12,000	101

cells generally have lower CO tolerance, hence, requiring higher H_2 purity than high temperature ones. Only 10–20 ppm CO is tolerable for LT-PEMFC and <1% for HT-PEMFC and PAFC, requiring extra purification processes [2,37]. MCFC and SOFC can stably operate in the atmosphere of CO or syngas, thus, hydrocarbon fuels can be directly fed into these two high temperature fuel cells via internal reforming. Generally, reforming hydrocarbon fuels requires a steam-to-carbon of over 2.5 to avoid carbon deposition.

In thermodynamics, some fuels have shown higher theoretical efficiency. For example, Fig. 3.1B shows fuel cells using carbon have a theoretical efficiency of ~100%, and the ones using CH_4 (major content of natural gas) have a theoretical efficiency of over 90%. However, either external or internal fuel processing increases the system complexity and lowers system efficiency due to temperature difference between fuel cells and fuel processing. These factors result that the fuel cell systems using these promising fuels may be less efficient. This encourages us to pay our efforts on developing novel fuel cells to realize a more compact integration with diverse fuels. Table 3.3 shows the thermodynamic features of typical fuels capable or potentially capable of being fed into fuel cells directly. Among these diverse fuels, hydrogen, and methane (main content of natural gas) have been commercialized in PEMFC or SOFC, respectively. The fuel cells directly fueled with other fuels are still under development or preparing for commercialization. However, other fuels are with high energy density by volume and easier to store or transport than H_2 or CH_4 due to liquid or solid phase, meanwhile, they all have theoretical efficiencies of over 95% when applied in fuel cells. Novel fuel cells directly using these fuels are expected to play much more significant roles in the future application scenes related to DHS and Energy Internet. Here, we give some brief introductions about some novel fuel cells, including direct liquid fuel cells, direct carbon fuel cells, and direct flame fuel cells.

3.2.6 Direct liquid fuel cells

Compared with hydrogen or other gaseous fuels, liquid fuels have higher energy density by volume and easier storage and transport. Direct liquid fuel cells (DLFCs) has provided high energy density, simple but compact structure, small fuel storage tank, and instant recharging, considered as one of novel and promising fuel cell technologies. DLFCs are capable of providing longer endurance for portable electronic devices such as cellular, and also extremely suitable for mid-size powers ranging from ~100 W to 3 kW, power supply in remote area and military applications (including field portable power sources, UAV, air-independent propulsion for submarines, armored cars, etc.) [39].

According to diverse liquid fuels, DLFCs involve direct methanol fuel cell (DMFC), direct ethanol fuel cell (DEFC), direct formic acid fuel cell (DFAFC), DAFC, direct hydrazine fuel cell (DHFC), etc. It should be mentioned that ammonia is easy to be liquefied (over 8 bar at 25°C) or dissolve in water in spite of gaseous phase at normal pressure and 25°C, therefore, sharing similar advantages such as high energy density by volume, easy storage, and transport with other liquid fuels. Consequently, we believe that it is reasonable to classify DAFC to DLFC. Comparison among typical fuels for DLFCs in Table 3.3 shows hydrazine (N_2H_4) and ammonia borane (NH_3BH_3) have the highest equilibrium potential E^0 (1.62 V) of all the listed liquid fuels, revealing that N_2H_4 and NH_3BH_3 are potential to generate more electricity per mole electrons transferred. For energy density, liquid fuels usually have lower energy density by weight but higher energy density by volume than hydrogen or natural gas (except formic acid). Ethanol and ammonia borane contain over 6000 Wh L^{-1} chemical energy, over fourfolds of that of H_2. Formic acid and ammonia both have a theoretical efficiency η_{th} of more than 100%, signifying a part of low-grade heat from environment could be converted into electricity. Table 3.4 summarizes current application fields and corresponding companies for different DLFCs.

TABLE 3.4 Application status of different DLFCs [39].

Type	Applied field	Examples
DMFC	Electronic devices	Fuel cartridge for laptop computer: Toshiba, LG, Panasonic, NEC, Fujitsu, Samsung Mobile phone charger: DoCoMo, Fujitsu Power pack: Toshiba, Panasonic
	Military equipment	Portable military device: Jenny 600, Jenny 1200 Onboard military vehicle generator: Emily DMFC generator
	Medical field	Yorkshire Ambulance Services: United Kingdom Hearing aid: Denmark
	Industrial field	Material handling equipment: Oorja Protonics
	Telecommunication	Backup power supply for telecommunication base station: IdaTech
	Security system	Police radio station, fire alarm system, smoke detector
	Educational kits	ENESSERE Horizon Bio Energy Education Kit
DEFC	Electronic devices	Power pack: NDCPower
	Home gadget	Vacuum cleaner: Bac Vac
	Automobile	Shell Eco-marathon DEFC prototype car
	Educational kits	Bio-energy Discovery Kit
DFAFC	Electronic devices	Fuel cartridge for laptop computer: Fuel Cell Research Centre, Republic of Korea
	Automobile	Unmanned aerial vehicle: Neah Power System and Silent Falcon
DHFC	Automobile	DHFC concept vehicle: Daihatsu Kei

Conventionally, a DLFC are likely to be based on an acid fuel cell (similar to PEMFC) or an alkaline fuel cell (similar to AFC) [39]. DMFCs are the most widely studied and the most established type among diverse DLFCs, which have been measured in Nafion membrane-based PEMFC [40] and AEM-based AFC [41]. DLFCs based on acid fuel cells share similar features with PEMFC. Acid DLFC electrolyte typically offers higher power density and faster start-up, but requires expensive noble-metal catalysts and has higher fuel crossover [39]. AFC-based DLFCs can replace noble-metal catalysts with nonnoble-metal ones and alleviate fuel crossover (due to ions conducting from cathode to anode), but sacrifice electrochemical performance and stability due to slow hydroxide ion conductivity and high CO_2 sensitivity, especially for carbonaceous liquid fuels.

Ref. [39] systematically reviewed current applications of different types of DLFCs, mainly involving electronic devices, military equipment, medical field, industrial field, telecommunication, security system, home gadget, automobile, educational kits, etc. As mentioned before, DMFCs have the widest application cases of all. The prototypes of DMFCs have been invented by a number of companies or institutes with the potential applications in the common electronic devices, military field, medical field, industrial field, telecommunication, security systems, and educational field. Toshiba and LG Chen Ltd. developed the prototypes of DMFCs with power outputs of 12 and 25 W to power a laptop continuously for over 5 and 10 h, respectively [39]. LG Chen Ltd.'s portable DMFCs are proven to have a lifetime as long as more than 4000 h. Similar portable DMFCs for laptops were also developed by NEC, Fujitsu, Panasonic, Samsung Electronics, etc. Panasonic developed an energy-efficient DMFC stack and its BOP system (including a fuel supply pump), which can regulate the methanol concentration fed into the stack [42]. This stack has been proven to offer a durability of over 5000 h (8 h intermittent supply per day) and an average power of 20 W. At a similar scale, NTT DoCoMo and Fujitsu Laboratories co-developed DMFCs fueled with 99% methanol for charging mobile phones [43]. Toshiba and Toyo Seikan Kaisha Ltd. co-developed a compact DMFC power pack, named as Dynario™, to generate a maximum power of 2 W with a fuel storage capacity of 14 mL and a total weight of 280 g and charge via a USB port [44]. SFC Energy invented a new portable DMFC, named Jenny 1200, for US Air Force to generate a rated power of 50 W with a maintenance-free period of 2500 h, hence to save up to 80% of carry-along weight for soldiers and extend their mission period by several days [45]. Besides, SFC Energy developed another military DMFC system with a power output of 0.5 kW for onboard armored vehicles, which have become the most successful military fuel cell system of SFC [46]. Recently, SFC Energy launched an original 6 kg (13 lb) fuel cell/battery power system named EFOY GO! for on-road transportation application, potentially offer a 25% higher capacity (25 Ah/300 Wh) than previous battery-based power system. DMFCs are also potential as the backup power of police radio station, fire alarm system, and more for guaranteeing security in the emergency situations [47]. Furthermore, DMFCs have been selected to power the equipment in the medical field. The Yorkshire Ambulance Service in United Kingdom uses DMFCs to supply reliable power for the medical equipment in an ambulance [48]. DMFCs with methanol volume of only 0.2 mL are used to power the hearing aids in Denmark, which can operate for over 24 h and offer a lifetime of over 6000 h [49]. For a kW-level scale, DMFC systems (1.5–4.5 kW) has been used for charging the battery of material handing vehicles produced by Oorja Protonics [39], hence, reducing battery capacities and increasing productivity.

DEFCs have developed for powering electronic devices, home gadget, automobile, and educational kits. NDCPower is the leading company of DEFC power pack, who has launched EOS DEFCs with power outputs varying from 1 to 250 W for powering laptop computers and mobile phones [39]. Similarly, DFAFCs have also been successfully used for powering laptop computers by Fuel Cell Research Centre, Republic of Korea [50]. DEFCs invented by BacVac show their ability to power vacuum cleaners and make them get rid of power wire. In a larger scale, a prototype DMFC-powered car was shown in Shell Eco-marathon Asia 2012 with a travel range of 2903 km per liter ethanol [51]. High efficiency and long travel range reveal their excellent prospect in the transport application. A novel application of DFAFCs in UAV has also been realized for commercial, public safety, and military applications by Neah Power and Silent Falcon UAS Technologies [52]. The formic acid-based fuel cell system, named Silent Falcon, could double or triple the mission endurance of UAV, and also increase the weight-carrying ability. Daihatsu Kei proposed a concept vehicle fueled hydrazine for cleaner and cheaper power supple due to the use of nonnoble metal catalyst (Cobalt or nickel) and carbon-free fuel [39]. Daihatsu has achieved a power density of 500 mW cm^{-2} from their DHFCs, making them comparable to H$_2$-fueled fuel cells. However, hydrazine is extremely toxic, which may need extra consideration to reduce the risk level when fuel leakage. Ammonia is also a carbon-free fuel with high energy density. Owning to its pungent odor, NH$_3$ detection is self-alarming. Just by smelling, it is possible to notice a leakage once NH$_3$ presence exceeds 5 ppm by volume [53]. Compared with hydrazine, NH$_3$ has lower toxicity. The Occupational Safety and Health Administration (OSHA) sets an acceptable exposure limit of 8 h at 25 ppm [54], while a NH$_3$ content of 5 ppm is well below any danger or damage. Recently, DAFCs make a new progress, revealing their competitiveness with other DLFCs. POCellTech realized a power density as high as 0.42 using an AEM-based DAFC [55]. A clean, safe, and cheap power solution may be realized by DAFCs in the future.

3.2.7 Direct carbon fuel cells

Coal, with a main component of carbon, is cheap, energy-dense, and easy to store. The amount of coal is abundant in nature, particularly in China. The volumetric energy density of coal is ~23.9 kWh L^{-1} [56], one order of magnitude higher than that of hydrogen (2.49 kWh L^{-1}) and 2.66 folds of that of gasoline

(9.0 kWh L^{-1}) [57]. Conventionally, carbon-to-electricity needs to suffer a series of energy conversion including carbon-to-heat through combustion, heat-to-mechanical energy through heat engine, and mechanical energy-to-electricity through electric generator. A direct carbon fuel cell (DCFC) is a novel electricity generation device, which can directly convert the chemical energy in carbon (rich in coal, petroleum coke, biomass, solid waste, etc. [58]) to electrical power. As Table 3.4 shows, DCFC has a theoretical efficiency of 101%. In DCFC, carbon and air are separately fed into anode and cathode. CO_2 or other contaminants from carbon/coal is not necessary to mix with N_2, thus, easier to capture. However, solid carbon-rich fuels have relatively slow reaction kinetics. This requires higher temperature and higher pressure to obtain an acceptable performance [57]. Currently, aqueous AFC at <250°C, molten hydroxide-based AFC at 500–600°C, MCFC at 750–800°C, and SOFC at 500–1000°C are under development to convert carbon to electricity directly [2].

Aqueous AFC used for DCFC requires an operating pressure as high as 30–35 bar to maintain an aqueous electrolyte solution below 250°C, which has been proved but is hard to obtain acceptable performance. Molten hydroxide-based AFC has been applied for DCFC to elevate the temperature to higher than 500°C. A 1 kW-level DCFC was successfully built using molten hydroxide-based AFC by William Jacques in 1896 [59]. Limited by carbonate formation and impurity contamination, this DCFC failed to commercialize. Scientific Applications and Research Associates (SARA) has developed four molten hydroxide-based DCFC demonstration including two latest prototypes, that is, MARK II-D and MARK III-A. MARK II-D uses Ni foam lined steel cathode to offer a maximum power density of 58 mW cm^{-2} and ∼100 h lifetime, while MARK III-A uses Fe$_2$Ti cathode to achieve 540 h lifetime with a maximum power density of 42 mW cm^{-2} [2].

MCFC is one of the most promising DCFC device, potentially converting near 100% carbon fuels and having an electric efficiency of ∼80%. Besides, MCFC also shows a series of advantages including long-term operation in CO_2 atmosphere, active carbon catalytic oxidation, and enough ionic conductivity. The compatibility of MCFC with diverse carbon-rich fuels have been studied, including acetylene black, coal derived carbon, furnace oil carbon black, graphite particles, heat treated petroleum coke, etc. [2]. Cooper et al. from Lawrence Livermore National Laboratory (LLNL) used furnace black, fossil chars, petroleum coke, bio-chars, and carbon blacks to obtain a power density of up to 100 mW cm^{-2} with an electric efficiency of 80% [60]. They invented a self-feeding subsystem and integrated it with a 5-cell stack to generate 75–150 W when fed. Presently, a suitable fuel delivery and processing system is still lacking for direct carbon-fueled MCFC and a long-term stability at kW-level scale still needs to be proved. However, the application of direct carbon-fueled MCFC has to address technical issues involving cathode degradation, corrosion of metal bipolar plates, fuel processing and delivery, reversed Boudouard reaction (should keep at a constant polarization voltage).

Typical direct carbon-fueled SOFC (DC-SOFC) uses oxygen-conducting solid oxide electrolyte (commonly YSZ), which has lower requirement for the impurity of carbon fuels. DC-SOFC is possible to integrate with existing coal burner such as fluidized bed, hence, offering a solution for fuel delivery. According to different anodes, DC-SOFC can be classified into porous-anode DC-SOFC, molten carbonate DC-SOFC, and molten metal DC-SOFC. Porous anode DC-SOFC uses the same anode structure as SOFC, where carbon can be directly oxidized, or gasified into CO for CO electrochemical oxidation. Direct carbon oxidation in porous anode requires triple-phase boundary (TPB) where carbon, ionic conductor, and electronic conductor co-exist. However, solid-solid contact lowers electrochemical kinetics because the consumption of carbon at TPB could create the gap between carbon fuel and TPB. Therefore, more efforts are paid to integrate carbon gasification with porous anode SOFC. Gür et al. [61] placed solid carbon and YSZ-based SOFC in two different temperature regions and studied the match between carbon gasification and SOFC in reaction rate. When SOFC operated at 800°C, carbon can offer enough gasification rate at 796–910°C. Gür's latter experiment obtained a power density as high as 450 mW cm^{-2} through the integration of an external Boudouard gasifier and SOFC [62]. A compacter integration of carbon gasification and SOFC needs to operate these two processes at the same temperature. Thus, carbon gasification needs to be accelerated to match with electrochemical oxidation reaction. Yu and Li et al. [63,64] used K/Ca/Ni to accelerate carbon catalytic gasification. They found K-promoted carbon gasification can improve power density from 30 to 190 mW cm^{-2} at 800°C. The concept of porous anode DC-SOFC stacks have been proposed by Clean Coal Energy and Akron University [2,65], separately, however, porous anode DC-SOFC still faces with a series of challenges. Sulfur containing in real carbonaceous fuels should be removed, or leads to the degradation of porous anode DC-SOFC. Fuel delivery and processing, device scale-up and thermal cycles are also currently unsolved. To promote direct carbon electrochemical oxidation, liquid anode SOFCs were invented to convert carbon contact from solid-solid contact to solid-liquid contact for faster electrochemical kinetics. Presently, two liquid anodes have been proposed, that is, molten carbonate anode and molten metal anode. Molten carbonate DC-SOFC is the combination of SOFC and MCFC. In anode, carbon fuels are mixed with molten Li_2CO_3 or K_2CO_3. Balachov et al. [66] at the SRI International tested tubular molten carbonate DC-SOFC fueled with diverse carbonaceous fuels including acetylene black, biomass, coal, tar, and mixed plastic waste. Acetylene black as anode fuel showed the highest power density (125 mW cm^{-2}) of all carbonaceous fuels, coal offered the second highest one (110 mW cm^{-2} at 950°C) and the biomass (NIST 8493 Pinus Radiata) showed the lowest one (only 60 mW cm^{-2} at 950°C). Based on this tubular-type molten carbonate DC-SOFC, SRI International developed a multiple tubular DCFC stack, where eight YSZ-electrolyte tubes inserted into a carbon fuel-mixed molten carbonate chamber [67]. This prototype has achieved a power density as high as 300 mW cm^{-2} and a lifetime of 1200 h. However, molten carbonate DC-SOFC has to address similar issues to MCFC, that is, the corrosion of current collectors, the stability of YSZ electrolyte-molten carbonate interface, molten carbonate circulation, slow electrochemical kinetics, etc. [57]. In molten metal DC-SOFC, molten metal participates the electrochemical reaction to form metal oxide, then carbon fuels reduce metal oxide back to molten metal and produce CO_2. Compared with molten carbonate, molten metal as SOFC anode can offer easier current collection, high impurity tolerance, even higher volumetric TPB density (due to ionic conductivity of some metal oxide, such as Sb_2O_3). Molten metal DC-SOFC can also operate in battery mode for continuous power generation even when fuels are lacking, hence, enables intermittent fuel delivery. Sb_2O_3 as molten metal oxide can realize 95% complete oxidization of carbon. CellTech Power has proposed a series of patents related to molten metal anode, and demonstrated tubular liquid tin anode SOFC (LTA-SOFC) system supported by DARPA (US Defence Advanced Research Projects Agency) [2]. Gür also applied a patent related to carbon-mixed molten metal bath (Ag and Bi) as SOFC anode [68]. CellTech Power has tested their LTA-SOFC system to obtain power density of 160 and 80 mW cm^{-2} when feeding H_2 and JP-8, respectively [69]. This system fueled with JP-8 has continuously operated for 200 h with an efficiency of >30%. Carbon, coal extracts, and biomass have been also tested in their single cell. Single cell test using carbon showed a fuel efficiency of 48%–62%, higher than using JP-8, while that using biomass only had a fuel efficiency of 34%. However, in LTA-SOFC system, carbon-mixed liquid tin was separated from anode using a porous ceramic separator, which causes extra diffusion resistance, particularly when feeding carbonaceous fuels. When liquid tin is used as SOFC anode, a significant degradation in electrochemical performance was found [70]. SnO_2 has a melting point as high as 1927°C and could be formed at TPB to block oxygen ion transport. To replace SOFC anode with molten metal, the melting point and ionic conductivity of molten metal oxide need to be considered as well. Therefore, Bi and Sb become promising alternative molten metal anode for DCFC.

Overall, DCFC is a promising technology with high efficiency, producing high-purity CO_2, and low fuel cost. Particularly in China, coal is still the major primary energy sources. An efficient and clean pathway for coal utilization is really in demand. However, DCFC is still at the early stage. There are some common challenges unsolved, including fuel processing and delivery, impurity sensitivity (typically S), slow carbon conversion, anode degradation, system scale-up and integration, etc. As a consequence, more efforts on R&D of DCFC are still needed to realize sufficient performance, enough lifetime and applicable scale.

3.2.8 Direct flame fuel cells

Conventional SOFCs have two separate chambers, that is, fuel-fed anode chamber and air-fed cathode chamber. However, the sealing of these two chambers at such a high temperature is an issue because of the mismatch in thermal expansion between SOFC and linking components. Direct flame fuel cell (DFFC) is a novel single-chamber fuel cell combining a fuel flame with SOFC, where anode is exposed in the fuel-rich flame region and cathode in air. DFFC has a series of distinct features including fuel flexibility, no need for sealing, simple configuration, and rapid start-up [71]. The concept of DFFC was proposed in Japan by Horiuchi et al. from Shinko Electric Industries [71]. In hydrocarbons-fueled DFFC, fuel-rich flame can offer heat and syngas-rich reactant flow for SOFC. On the one hand, DFFC is thermally self-sustainable. On the other hand, hydrocarbon fuels are partially oxidized to small-molecule fuels in fuel-rich flame and beneficial for continuous and stable operation of SOFC. Therefore, DFFC co-generates electricity and high-grade heat, which is extremely suitable for hybrid distributed systems, that is, CHP and CCHP. Heat generation from flame is fast enough for SOFC to realize second-level start-up, meanwhile, such fast start-up requires high thermal shock resistance of SOFC. Currently, DFFC is only demonstrated in laboratory scale at the R&D stage.

Original concept of DFFC proposed by Horicachi used Bunsen burner to burn *n*-butane, kerosene, paraffin wax, wood, etc. [71]. Nevertheless, cone flame from Bunsen burner causes a large spatial variation in temperature and gas composition, and further increases thermal stress of SOFC and enhances the mechanical failure possibility. Vogler et al. [72] used McKenna flat-flame burner to offer more uniform temperature and gas composition fields. They studied the effect of the equivalence ratio Φ and found that the peak power density (120 mW cm^{-2}) was obtained when $\Phi = 1.3$. But the mismatch between fuel flow rate and SOFC led to an electrical efficiency of only 0.45%. Wang et al. [73] developed a DFFC system, where they exposed tubular SOFC in the methane-rich flame produced by a porous media burner. This system can realize 10 s start-up and the OCV stabilized at ~0.9 V. They found an increase in the equivalence ratio from 1.3 to 1.6 can improve the power density at 0.65 V from 700 to 1200 mW cm^{-2}. Based on Wang's experiments, they designed a 14 microtubular SOFC stack with a burner-volume-specific power density as high as 370 mW cm^{-3}.

To realize the commercialization of DFFC, there is still a series of challenged to be solved, such as thermal shock and carbon deposition. Rapid start-up causes large temperature variation with time, leading to high thermal shock, which limits the use of planar-type SOFC. Microtubular SOFCs can offer higher mechanical strength and (many orders of magnitude) larger thermal shock resistance than planar ones [74]. Another serious issue of DFFC is carbon deposition. The partial oxidation of hydrocarbon fuels could produce apparent carbon and further cause anode pore blocking, performance degradation, mechanical failure, etc. Some fuel types and anode materials have been suggested to be suitable for DFFC, such as using methanol as fuel, and adding Ru-based catalytic layer [74].

3.3 Power-to-gas or power-to-liquid for energy storage

As evaluated, DHS needs approximately 15%–20% of the annual loads to meet 2–3 month of storage [75,76]. Other than pumped-hydro storage (PHS) and compressed-air energy storage (CAES), storing energy in fuels is the only energy storage technology applicable for seasonal energy storage [75,77]. Power-to-gas (PtG) or power-to-liquid (PtL) technology can convert surplus renewable power into gaseous or liquid fuels through electrolyzers and then store them with zero self-discharge losses. General conversion pathways and applications of PtG and PtL are shown in Fig. 3.4. The most common electrolyzers are water electrolyzers, where H_2O is electrochemically reduced to H_2. High-purity H_2 produced by electrolyzers can directly fuel cells and hydrogen power plants, or be transported to hydrogen refueling station. Essentially, PtG or PtL is to store renewable energy in hydrogen-rich fuels or chemicals. Even though power-to-hydrogen is the most common PtG pathway, hydrogen storage is still a challenge. High diffusibility, buoyancy, and activity of H_2 make it difficult to store [33]. Generally, H_2 needs to be pressurized to 350–700 bar for storage and transportation, which causes an increase of risk level. Some hydrogen storage technologies such as cryo-compression, metal-hydrides and adsorption are in development, but still not cost-efficient. Storing hydrogen in chemical hydrides reveals lower costs and higher technical maturity than other hydrogen storage technologies. Hydrocarbons such as methane, methanol, and gasoline are usually selected as the storage media, which can also recycle CO_2 emitted from fossil fuel-based power plants or chemical industries. CO_2 can react with H_2 generated from water electrolyzers to produce various hydrocarbons, or CO_2 and H_2O are co-electrolyzed to syngas ($H_2 + CO$ mixture) for hydrocarbon production even directly to produce target hydrocarbons. Hydrocarbons have wider applications including power generation, refueling station, multiple chemicals, and jet fuels. Particularly, liquid hydrocarbons are more convenient for transportation, hence, can be easier to be sold to downstream markets. Power-to-ammonia (PtNH$_3$) is a novel route for PtG or PtL, which is the only carbon-free conversion pathway among these mentioned ones in Fig. 3.4 except power-to-hydrogen. Conventionally, NH_3 is an important chemical for fertilizer industry, refrigeration, manufacture

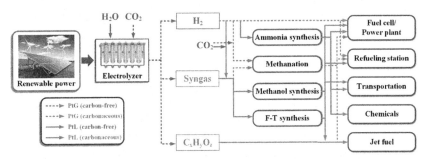

FIGURE 3.4 General conversion pathways and applications of PtG and PtL.

(plastics, explosives, textiles, pesticides, dyes, and other chemicals), cleaner, etc. In fact, NH_3 is attracting more and more attentions as a good H-rich energy carrier due to high energy density, carbon-free, easy liquidation, and low flammability.

3.3.1 Electrolyzers

Typical electrolyzers including alkaline electrolysis cells (AECs), proton exchange membrane electrolysis cells (PEMECs), and solid oxide electrolysis cells (SOECs). The half reactions, electrolyte materials, and operating temperature for these three types of electrolyzers are summarized in Table 3.5. AEC and PEMEC generally operate below 100°C, belonging to low-temperature electrolyzers. The operating temperature of SOEC is much higher than the former two electrolyzers, that is, 400–1000°C.

3.3.1.1 Energy demand and efficiency of electrolysis cells

Thermodynamics can judge whether a reaction happens spontaneously or not, and calculate the minimum energy required to sustain the electrolysis process. In fact, OCV denotes the minimum electricity demand, which can be predicted by the reversible Nernst potential E_N. Electrolysis is the reversed process of fuel cell, thus, Nernst potential E_N has the following relation with the Gibbs free energy change ΔG of the electrolysis based on Eq. (3):

$$\Delta G = n_e F E_N \tag{7}$$

Considering the effect of temperature and gas composition, E_N can be expressed as [78]:

$$E_N = \frac{\Delta G^0}{n_e F} - \frac{\Delta S}{n_e F}(T - 298.15) + \frac{RT}{n_e F} \ln \frac{\prod p_{products}^{v_i}}{\prod p_{reac\tan ts}^{v_i}} \tag{8}$$

TABLE 3.5 Typical electrolyzers under development (classified by electrolyte types).

Electrolyzer type	Cathode reaction	Electrolyte	Anode reaction	Temp. (°C)
AEC	$2H_2O + 2e^- \rightarrow H_2 + 2OH^-$	$Me(OH)_x$ $OH^- \rightarrow$	$2OH^- \rightarrow 1/2O_2 + H_2O + 2e^-$	<100
PEMEC	$2H^+ + 2e^- \rightarrow H_2$	PEM H^+	$H_2O \rightarrow 1/2O_2 + 2H^+ + 2e^-$	60–100
SOEC	$H_2O + 2e^- \rightarrow H_2 + O^{2-}/2H^+ + 2e^- \rightarrow H_2$	Ceramic $O^{2-}/H^+ \rightarrow$	$O^{2-} \rightarrow 1/2O_2 + 2e^-/H_2O \rightarrow 1/2O_2 + 2H^+ + 2e^-$	600–1000 (O-SOFC) 400–800 (H-SOFC)

where, ΔG^0 is the Gibbs free energy change in standard condition (25°C, 1 atm), ΔS is the entropy change, and v_i denotes the stoichiometric coefficients in the overall reaction. For instance, E_N for H_2O electrolysis is:

$$E_N = \frac{\Delta G^0}{n_e F} - \frac{\Delta S}{n_e F}(T - 298.15) + \frac{RT}{n_e F} \ln \frac{p_{H_2} p_{O_2}^{1/2}}{p_{H_2O}} \qquad (9)$$

In an isothermal-isobaric process, the enthalpy change of electrolysis reaction is the theoretical total energy demand. The enthalpy change has the following relation with the Gibbs free energy change and the entropy change:

$$\Delta H = \Delta G + T \Delta S \qquad (10)$$

Thus, $T\Delta S$ is the minimum heat demand for the electrolysis process. An electrolysis process demands at least electrical energy input (ΔG) and heat energy input ($T\Delta S$). Fig. 3.5 shows the energy demand for both H_2O electrolysis and CO_2 electrolysis at 0–1000°C. At <100°C, H_2O electrolysis requires higher energy demand (both electrical demand and heat demand) due to liquid H_2O. For gaseous H_2O, temperature rise causes the required electricity decreasing and the required heat increasing, while the total energy demand varies little. H_2O electrolysis requires at least 237 kJ electricity at 25°C and at least 189 kJ electricity at 800°C. Consequently, high temperature electrolysis using SOECs potentially saves over 25% electricity consumption. As for the rise in heat demand, some high-grade waste heat from the DHS, like the waste heat from power plants, chemical industries, etc., can be recovered to promote the overall efficiency of DHS.

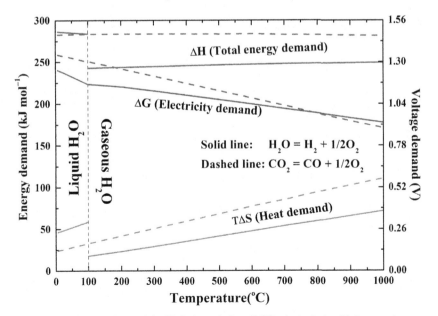

FIGURE 3.5 **Energy demand for H_2O electrolysis and CO_2 electrolysis with temperature.**

In real operation, electrolyzers inevitably need to overcome the irreversible polarization losses or ohmic losses. Thus, electrolyzers operate at a voltage higher than OCV, opposite to fuel cells. The difference between operating voltage and OCV is the polarization voltage η. The irreversible losses convert a part of electrical energy input to heat. The released heat can supply for heat demand of electrolysis process. Higher the operating voltage is, higher the released heat is. When the released heat ($\eta n F$) is exactly equal to the heat demand of electrolysis process ($T\Delta S$), the corresponding operating voltage (V_{cell}) is called thermos-neutral voltage (TNV). Theoretically, TNV can be calculated as:

$$V_{TN} = V_{OC} + \eta = \frac{\Delta G}{nF} + \frac{T\Delta S}{nF} = \frac{\Delta H}{nF} \tag{11}$$

Typically, the TNV is 1.28 V for H_2O electrolysis and 1.47 V for CO_2 electrolysis. Generally, electrolyzers operate above TNV. Therefore, the thermodynamic efficiency η_{th} for electrolyzers is equal to the ratio of the high heating value (HHV) of produced fuel to the minimum electrical energy input W_{el}. Here, the efficiencies are considered in an isothermal system, hence, heat input is not considered. For electrolysis, HHV of produced fuel is equal to the value of the enthalpy change ΔH. Therefore, η_{th} can be expressed as:

$$\eta_{th} = \frac{HHV_{Fuel}}{W_{el,min}} = \frac{\Delta H}{\Delta G} = \frac{V_{TN}}{E_N} \tag{12}$$

For H_2O electrolysis or CO_2 electrolysis, the value of η_{th} is larger than 100% due to extra heat demand. Considering the irreversible losses, the voltage efficiency η_V is expressed below:

$$\eta_V = \frac{W_{el,min}}{W_{cell}} = \frac{E_N}{V_{cell}} \tag{13}$$

Consequently, the overall electrical efficiency of an electrolyzer η_{el} can be calculated as the following equation:

$$\eta_{el} = \eta_V \eta_{th} = \frac{V_{TN}}{V_{cell}} \tag{14}$$

Theoretically, η_{el} can reach 100% when the electrolyzer operates at TNV. The main features for these three electrolyzers are summarized in Table 3.6 [79]. In the following parts in this section, these features will be mentioned and discussed.

3.3.1.2 Alkaline electrolysis cells

AEC, the reversed process of AFC, is the first proposed electrolyzer among all the electrolyzers. The electrolyte of AEC is 25–30 wt.% KOH or NaOH solution. Currently, AEC is the most developed electrolyzer and also has the

TABLE 3.6 Main features of AEC, PEMEC, and SOEC [79].

Features	AEC	PEMEC	SOEC
Cathode	Ni/Ni-Mo	Pt/Pt-Pd	Ni-YSZ/Ni-ceramic
Anode	Ni/Ni-Co	RuO_2/IrO_2	LSM-YSZ
Current density (mA cm^{-2})	200–400	600–2000	300–2000
Operating voltage (V)	1.8–2.4	1.8–2.2	0.7–1.5
Electrical efficiency (%)	62–82	67–82	<110
Operating pressure (bar)	<30	<200	<25
Scale (Nm$^3_{H2}$ h^{-1})	<760	<40	<40
Stack electricity consumption (kWh Nm^{-3})	4.2–5.9	4.2–5.5	>3.2
System electricity consumption (kWh Nm^{-3})	4.5–6.6	4.2–6.6	>3.7
Lower dynamic range (%)	10–40	0–10	>30
System response	Seconds	Milliseconds	Seconds
Cold-start time (min)	<60	<20	<60
Stack lifetime (h)	60,000–90,000	20,000–60,000	<10,000
Maturity	Mature	Commercial	Demonstration
Capital cost (€ kW$_{el}^{-1}$)	1000–1200	1860–2320	>2000

lowest capital cost (1000–5000 $/kW varying with scale [80]). The stack lifetime of AEC can reach as long as 60,000–90,000 h (7–10 years). However, AEC has similar issues to AFC, such as the corrosion and leakage of aqueous electrolyte, slow electrochemical kinetics and high polarization losses (high electricity consumption). Fig. 3.6 shows the comparison among three types of electrolyzers. Currently, the maximum operable current density of AEC is around 400 mA cm^{-2}, only 20% of that of PEMEC or SOEC. Slower electrochemical reaction rate of AEC indicates that larger effective reaction area (larger volume) is needed than PEMEC and SOEC for the same hydrogen production rate. AEC generally operates at 1.8–2.4 V, higher than PEMEC and SOEC. This means that electricity consumption of AEC is ~19% higher than that of PEMFC, and 46% higher than that of SOEC for the same hydrogen production. The electrical efficiency of AEC is 62%–82%, and an AEC system generally requires 4.5–6.6 kWh electricity per Nm^3 H$_2$ production. Therefore, current studies on AEC mainly focus on enhancing electrochemical performance, pressurizing, and dynamic operation for addressing intermittent renewable energy.

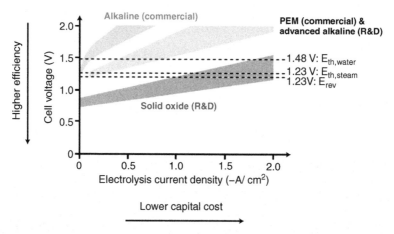

FIGURE 3.6 Typical polarization curves of PEMEC, AEC, and SOEC. *(Modified from Ref. [81] Copyright 2010 Elsevier Ltd.)*

The main manufacturers of AEC mainly include AccaGen, Avalence, Claind ELT, Erredue, Hydrogen Technologies, Hydrogenics, H2 Logic, Idroenergy, Industrie Haute Technologie, Linde, PIEL, Sagim, Teledyne Energy Systems, etc. [80]. Among these manufacturers, most of them use the bipolar modules. Avalence and Sagim are the representations of the monopolar cell manufacturers. AEC is the most suitable for large-scale hydrogen production, typically in the scale of 500–760 Nm^3 h^{-1} (electricity consumption 2100–3500 kW) [80]. AEC has a load regulation range from 25% to 100% and a second-level load response.

3.3.1.3   Proton exchange membrane electrolysis cells

PEMEC is the reversed process of PEMFC, based on a PFSA polymer electrolyte (the most common one is Nafion from DuPont). The use of nonaqueous electrolyte enables PEMEC to have compacter structure and higher electrochemical performance (than AEC). The current density for PEMEC can reach ~2000 mA cm^{-2} and the electrical efficiency is above 67%. The electricity consumption of PEMEC systems is about 4.2–6.6 kWh for 1 Nm^3 H_2. PFSA polymer membrane has low gas crossover rate, hence, PEMEC has fast response to load changes (millisecond-level) and almost 100% turndown capability. Therefore, PEMEC is quite suitable for renewable energy applications. Furthermore, PEMEC has no issues related to corrosion and gas sealing. Therefore, PEMEC can operate at higher pressure to compact the electrolysis system. The tolerance for differential pressure between two electrodes has higher safety level, and also allows to utilize ambient pressure oxygen in the oxygen/fluids loop to lower capital cost [82]. PEMEC operating at 350 bar is more attractive because such a high pressure may remove hydrogen compression from this system. Proton Energy System has realized a 20,000 h operation of PEMEC at 165 bar and also designed a 350 bar fueling concept for the Department of Energy [82]. However, the acidic membrane limits the use of nonnoble metal catalysts. The need for noble-metal catalysts (Pt-based for cathode, Ru- or Ir-based for anode) leads to higher capital cost. High impurity sensitivity of PFSA membrane also requires higher H_2O purity.

PEMEC has been commercialized, but the companies for PEMEC manufacture are fewer than those for AEC. Majority of them are from USA and Canada, including Giner, Hydrogenics, Proton Onsite, Treadwell Corporation[80]. PEMEC is not as mature as AEC, hence, mainly applied in some smaller-scale applications. The maximum scale for PEMEC is $\sim 30 \, Nm^3 \, h^{-1}$ (electricity consumption 174 kW) [80]. The capital cost for PEMEC is approximately 1800–2300 € per kW_{el}.

3.3.1.4   Solid oxide electrolysis cells

SOEC is a high temperature electrolyzer utilizing steam instead of liquid water. SOEC is the reversed process of SOFC, similarly, the common electrolyte of SOEC is also YSZ. Elevated temperature enables SOEC to operate at a much lower operating voltage (0.7–1.5 V). According to Section 4.1.1, the electrical efficiency significantly improves, even exceeds 100%. Therefore, the electricity consumption for SOEC can reduce to lower than 4 kWh per Nm^3 H_2. Particularly, some power plant, chemical industries, and nuclear plants with high-grade waste heat sources shows high compatibility with SOEC systems. Besides, some renewable heat sources, such as solar energy and geothermal energy, can be utilized more efficiently when integrated with SOEC systems. However, SOEC is the least developed electrolyzers among these three electrolyzers, which is still at the early stage of commercialization. The major issues of SOECs are related to the stability of materials and gas sealing between gas channels. Most of SOEC stacks are still hard to offer enough lifetime (typically <10,000 h). Therefore, SOEC systems have a capical cost of >2000 € per kW_{el} at current stage. Nevertheless, the materials of SOEC like catalysts and electrolyte are not expensive, thereby, the capital cost of SOEC is expected to reduce dramatically with the maturation and commercialization of this technology. The companies who are active in the demonstration and commercialization of SOEC systems include early Westinghouse Electric Co.,Toshiba Corporation, Ceramatec Inc., Dornier System GmbH, Haldor Topsoe, etc.

High temperature operation makes SOEC with highly CO-tolerance. Even SOEC can directly electrolyze pure CO_2 to CO. Haldor Topsoe proposed an electrolytic carbon monoxide solution called eCOs™ unit, where CO is produced from CO_2 as the energy carrier [83]. However, CO_2 electrolysis inevitably consumes more electricity than H_2O electrolysis due to stable C=C bond. Laboratory study on SOECs reveals that the electrochemical performance of CO_2 electrolysis is only 33%–50% of that of H_2O electrolysis [24]. To utilize CO_2 at a higher efficiency, co-electrolysis of H_2O and CO_2 using SOEC was proposed. Co-electrolysis of H_2O and CO_2 has an electrochemical performance close to H_2O electrolysis, meanwhile, is capable of converting H_2O and CO_2 to syngas with adjustable H/C ratio, which is ideal sources for methanation, methanol production, and F-T synthesis. Therefore, if co-electrolysis of H_2O and CO_2 is integrated with these mature chemical engineering processes, PtG or PtL has much higher flexibility in fuel product. Moreover, when SOEC is pressurized, it is possible to combine H_2O/CO_2 co-electrolysis and methanation reactor/methanol synthesis/F-T reactor into one reactor, that is, direct power-to-hydrocarbons. In such a compact reactor, the heat released by hydrocarbon synthesis can supply for electrolysis processes to realize an efficient thermal coupling [84].

3.3.2 Power-to-gas

In islanded DHS (such as those in remote area, small islands, or other special geographical locations), the fuel transportation and supply could be difficult or costly. Introducing renewable energy into islanded DHS is a feasible pathway. Intermittent renewable energy can be stored seasonally using electrolysis-based fuel production technology (PtG or PtL). Conventionally, DHS uses engines or gas turbines to convert the stored fuels back to electricity. The commercialization of fuel cells offers more efficient and cleaner options to utilize the stored fuels. Therefore, such a DHS generally requires an electrolyzer for fuel generation and a fuel cell for electrical power generation from the stored fuels, in which hydrogen is the most common storage fuel.

3.3.2.1 Power-to-hydrogen

The schematic of this PtG energy storage system is shown in Fig. 3.7. The DHS can store intermittent renewable energy and generate stable electricity in the meantime, hence, has the function of power supply stabilization.

The very first PtG systems utilize renewable power to convert H_2O to H_2, and the generated H_2 is stored in pressurized cylinder, metal hydride storage, or directly sent to gas pipelines. A study performed by National Renewable Energy Laboratory (NREL) reveals that less than 5–15 vol.% of H_2 in natural gas pipeline appears to be feasible without dramatically increasing risks for end-use devices [86]. An early review on PtG plants summarized the installed capacity of power-to-hydrogen (PtH_2) driven by renewable energy as Fig. 3.8 shows [87]. This review mentioned 41 realized PtH_2 plants and 7 planned ones with a wide installed capacity range from 1 kW_{el} to 6.3 MW_{el}, and most of them are from Germany, USA, Canada, Spain, and UK. Four PtH_2 systems for renewable energy storage have been realized in 1991–93. Hychico project in Argentina has the largest installed capacities (two 320 kW_{el} AEC) among all mentioned PtH_2 systems in operation [87]. The largest planned PtH_2 projects with a total installed AEC capacity of 6.3 MW_{el} are in Germany. Most of these PtH_2 plants are applied to offer electricity for local grid or public grid using fuel cells (the

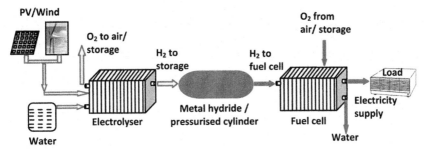

FIGURE 3.7   Schematic of a conventional PtG system for renewable energy storage. *(Modified from Ref. [85] Copyright 2017 Elsevier Ltd.)*

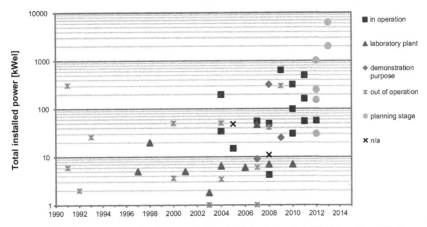

FIGURE 3.8   Installed capacity for realized PtH$_2$ pilot plants. *(Modified from Ref.[87] Copyright 2012 Elsevier Ltd.)*

majority selected PEMFCs and a few AFCs), internal combustion engines or CHP plants. MYRTE project in Corsica, France has the largest PEMFC capacity (100 kW$_{el}$) among these mentioned projects, and the other ones mainly installed fuel cells with the capacity of <10 kW$_{el}$ [87].

3.3.2.2   Power-to-methane

Despite a wide range of efforts have been paid on developing hydrogen-based technologies for establishing a clean and sustainable energy structure. However, hydrogen infrastructure and market are still not well established at current stage due to the cost, safety, and technical maturity of hydrogen-based technologies [88]. Thus, other alternative storage fuels are proposed, such as CH$_4$ to have high compatibility with the existing natural gas infrastructure. Natural gas (typically containing >95 vol.% CH$_4$) is applicable for heating, transportation, power generation, chemical industries, etc. [89]. Power-to-CH$_4$ (PtM) offers a more practical solution for the synchronous storage of renewable energy sources and carbon dioxide due to the wide distribution of natural gas pipelines. CO_2, as a significant feedstock of PtM, can be captured from the exhaust gas of coal power plants or coal chemical plants, or recycled from the integrated CH_4-based power generators.

Fig. 3.9 shows the diagram schematics for three different PtM pathways utilizing renewable power. The conventional conversion pathway is to integrate a water electrolyzer with a Sabatier reactor (SR: $CO_2 + 4H_2 = CH_4 + 2H_2O$) as shown as Route 1 in Fig. 3.9. Fig. 3.10 shows current PtG projects distributions by countries and applications. Germany is still the most active country to con-

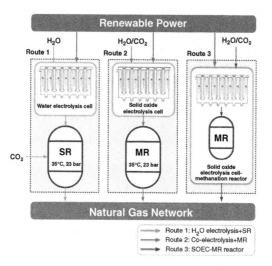

FIGURE 3.9 **Three pathways for PtM systems driven by renewable energy sources.** *(Modified from Ref. [84] Copyright 2018 Elsevier Ltd.)*

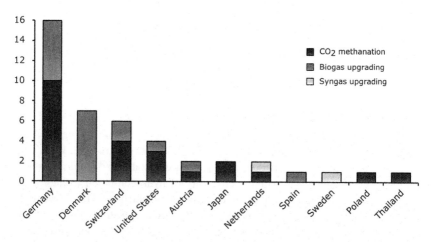

FIGURE 3.10 **Distribution of current PtG projects.** *(Modified from Ref. [88] Copyright 2016 Elsevier Ltd.)*

struct PtG plants, among which PtM projects account for over 60%. As reported, Germany is capable of offering a natural gas capacity of over 200 TWh based on the natural gas network and underground storage [90]. The first industry-scale PtM plant with the installed AEC capacity of 6.3 MW_{el} has been demonstrated in Werlte by ETOGAS for Auti AG in 2013 [88]. The largest PtM plant worldwide utilizes intermittent wind power and CO_2 feedstocks from a biogas plant and realizes a system efficiency of ~54%. Nevertheless, an efficiency loss as high as 16% is inevitably caused compared with PtH_2 plant because of additional CH_4 production process [89].

When using high temperature SOECs, PtM systems enable to convert CO_2 directly. Route 2 in Fig. 3.9 denotes a PtM pathway based on H_2O/CO_2 co-electrolysis using SOECs. In Route 2, SOEC produces syngas with adjustable H/C ratio for a sequent methanation reactor (MR, $CO_2 + 4H_2 = CH_4 + 2H_2O$ and $CO + 3H_2 = CH_4 + H_2O$). As operating pressure of SOECs rises and operating temperature of SOECs decreases, SOEC reactor shows the potential to realize direct H_2O/CO_2-to-CH_4 conversion as Route 3 reveals in Fig. 3.9. Our group has successfully built a pressurized tubular SOEC test-setup to realize a CO_2 conversion of 40% and a CH_4 yield of 28% at 650°C. Our group collaborating with Ghoniem's group from MIT, USA performed a comprehensive exergy analysis to evaluate the exergy efficiency of the above three PtM systems (Route 1–3) [84]. In Route 1, three types of electrolyzers,that is, AEC, PEMEC, and SOEC, are integrated respectively for comparison. Because of high operating temperature, SOEC-based system is less efficient than PEMEC-based system at a current density of lower than 3400 A m^{-2} (operating voltage is below TNV). As current density rises, SOEC-based system exceeds PEMEC-based one and reaches the peak efficiency (70%) at 8000 A m^{-2}, which is 11% higher than PEMEC-based system. The AEC-based system has the lowest exergy efficiency among these three systems, only less than 42%. The distributions of energy consumption show over 86% of energy consumption is used for driving the electrolyzers. This ratio in the systems using low-temperature electrolyzers reaches over 93%, 7% higher than SOEC-based system. In Route 2, the intermediate-temperature SOEC-based PtM system can run more efficiently and achieve a more compact design, particularly at elevated pressure. The optimal operating condition for SOEC-based PtM system was found at 6000 A m^{-2}, 650°C and with an H/C ratio of 10.54. Intermediate-temperature SOEC can promote thermal integration and improve exergy efficiency to over 77% at over 6 bar. In Route 3, the temperature mismatch of SOEC and methanation limits an isothermal SOEC reactor to produce high-purity CH_4 (at most 43.40% at 550°C and 25 bar). Therefore, our group studied an SOEC reactor with a gradient temperature field, revealing a great potential to improving efficiency. At <19 bar, the temperature-gradient SOEC reactor potentially performs at a higher efficiency than Route 2 (H_2O/CO_2 co-electrolysis + MR reactor). The optimal operating pressure was 8.15 bar, at which Route 3 has an exergy efficiency of 80%. In real operation, such an SOEC reactor should be optimized to realize effective heat transfer of fuel and air streams, heat recovery from the outlet gas and adequate methanation activity in cathode [84].

As the increasing PtG systems are installed, more topics should be noticed including the optimization of system configuration, the performance improvement of electrolyzers and fuel cells (related to efficiency, reliability, lifetime, maintenance, capital cost), the integration with intermittent renewable energy sources, BOP development, codes and standards for operating permission, control system and operating safety, etc. [87].

3.3.3 Power-to-liquid

The main feedstocks for PtL are electricity, H_2O, and CO_2. Similarly to PtM, conventional PtL system first electrolyzes H_2O to H_2, and then the produced H_2 reacts with CO_2 to produce target hydrocarbon fuels such as methanol, ethanol, and gasoline, as Fig. 3.11A shows. When using SOEC, PtL system can also co-electrolyze H_2O and CO_2 to produce syngas, and then syngas can be directly used as the feedstock of liquid fuel synthesis, corresponding to the schematic in Fig. 3.11B. There are two main technologies for liquid fuel synthesis, that

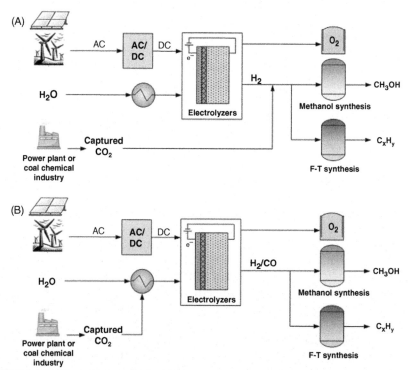

FIGURE 3.11 Typical schematics for PtL systems through H_2O electrolysis (A) and H_2O/CO_2 co-electrolysis (B).

is, methanol synthesis and Fischer-Tropsch synthesis. Up to now, there are still no commercial plants for power-to-jet fuels worldwide, and only several PtL system integrations are progressing in Iceland, Finland, Germany, and Norway [91]. However, the key technologies for PtL have already achieved sufficient technical readiness for industrial scale-up [91]. Sunfire has successfully operated a SOEC-based PtL demonstration plant in Dresden to produce 3 tons of Blue Crude over a continuous operating time of 1500 h. Sunfire is planning to launch a collaboration with Nordic Blue Crude, Climeworks, and EDL Anlagenbau to built a PtL plant with capability of producing 10 million liters per year in Norway [92].

3.4   Reversible fuel cells

Now that fuel cell and electrolysis are the reversed processes of each other, combining these two processes in one compact reversible reactor is possible, which is known as reversible fuel cells (RFCs). RFC can operate either in fuel cell (FC) mode or in electrolysis cell (EC) mode, hence, reduces the system volume or mass and saves the capital cost. When RFC is integrated with fuel storage, such a system is similar to a battery system without any self-charging losses.

There are two types of RFCs according to how the operating mode switches over. One type keeps the positive electrode (oxygen electrode) and negative electrode (fuel electrode) constant, that is, the half reactions in the electrodes occur reversely when operating mode switches over. The other one keeps the anode and cathode constant, that is, the current direction doesn't change but oxygen electrode and fuel electrode switch over. Most studies select the former type of RFC due to easier operation. In the former type of RFC, the product in EC mode can be fed back to the same electrode in FC mode. Expect for the gas stream management, the only operation needs to do for mode switchover is to regulate the operating voltage. However, a challenge for reversible operation is that conventional catalysts are hard to perform well in both directions of an electrochemical reaction [93,94], particularly for low-temperature RFCs. PEMFC generally uses carbon-based oxygen electrode (cathode), which can be corroded in the reversed reaction, that is, oxygen evolution in EC mode [85]. Therefore, reversible PEMFC (RPEMFC) has to make a trade-off to use metal-based oxygen electrode. Corresponding to Fig. 3.7, Fig. 3.12 shows the schematic of an RFC-based PtG system for renewable energy storage. RFC is integrated with a water tank and hydrogen storage component (pressurized cylinder, metal hydride storage, etc.). In EC mode, water is fed into RFC and electrolyzed into H_2 and O_2 driven by renewable energy. H_2 is stored in a pressurized cylinder or by other hydrogen storage technologies. In FC mode, the stored H_2 is fed back to RFC and reacts with O_2/air to generate electricity and the produced water is stored in the water tank. Therefore, H_2 and H_2O are cycled in the RFC-based PtG system.

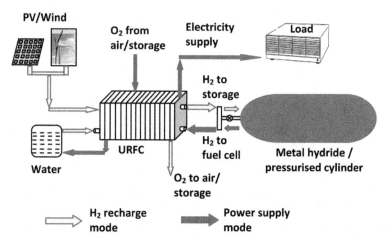

FIGURE 3.12 Schematic of an RFC-based PtG system for renewable energy storage. *(Modified from Ref. [85] Copyright 2017 Elsevier Ltd.)*

According to the electrolyte materials, main RFCs can be categorized to three types: reversible AFC (RAFC), RPEMFC, and reversible SOFC (RSOFC) (Fig. 3.13). These types of RFC correspond to the types of fuel cells and electrolyzers mentioned in Sections 2 and 3.1. These RFCs are compared in Table 3.7. RPEMFC is the most mature RFC among these three types of RFCs. The concept of RPEMFC was proposed by General Electric in 1973 for the application in space satellites. Until 1990s, LLNL, Proton Energy System Inc., United Technologies Corporation, Hamilton Standard, etc., began to apply RPEMFC systems for space satellites, zero-emission terrestrial vehicles, utility, and remote power energy storage, uninterruptible power supplies [85]. In 2016, a cooperation study carried out by National Institute of Advanced Industrial Science and Technology and Takasago Engineering Co. Ltd. demonstrated a PtG system with a 10-cells RPEMFC stack and a metal hydride subsystem [96]. This system can produce 2.3 kg hydrogen per day with an electricity consumption of 4.4 kW in EC mode and is capable of generating 0.68 kW DC power in FC mode. Besides, South China University of Technology, Japan Aerospace Exploration Agency, Hydrogen Energy and Plasma Technology Institute from Russia, etc., all demonstrated their respective RPEMFC prototypes with an electricity consumption of 0.1 W to 4.4 kW and an electricity generation of 0.07–680 W [85]. The scale-up of RPEMFC systems (kW level) is an important target at the nest stage of RPEMFC technology.

Differently RAFC and RPEMFC, RSOFC has higher operating temperature and is capable of utilizing CO_2 directly. Therefore, H_2 is not the only alternative energy carrier. CO_2 can be cycled in an RSOFC-based system, hence, a close-cycle between CO_2 and carbonaceous fuel can be formed. When using H_2 as the stored energy carrier, the round-trip efficiency of RSOFC is approximately

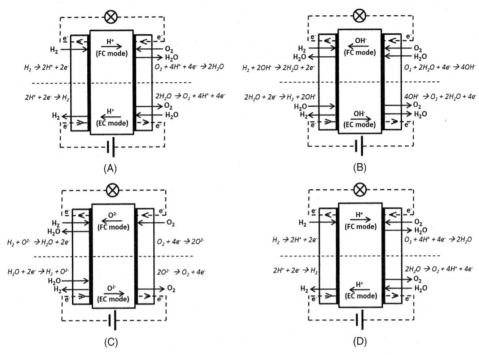

FIGURE 3.13   Working principles for RPEMFC, RAFC, oxygen-conducting RSOFC, and proton-conducting RSOFC. *(Modified from Ref. [95] Copyright 2016 Elsevier Ltd.)*

TABLE 3.7 Comparison among different RFCs.

Types	Main features	Temp.	Fuels	Round-trip efficiency [95]	Technical status
RAFC	Nonnoble-metal catalyst	20–120°C	H_2	30%–40%	Developing
RPEMFC	Noble-metal catalyst	20–100°C	H_2	40%–50%	Demonstration
RSOFC	High-temperature operation High round-trip efficiency High fuel flexibility	400–1000°C	H_2/CO/ CH_4	60%–80%	Developing

60%–80%, 20%30% higher than low-temperature RFCs. Theoretically, the round-trip efficiency η_{cyc} is the product of the electrical efficiency in FC mode (Eq. (6)) and in EC mode (Eq. (14)), which can be expressed as:

$$\eta_{cyc} = \eta_{el}^{FC} \eta_{el}^{EC} = \frac{nFV_{cell}^{FC}}{\Delta H} \frac{\Delta H}{nFV_{cell}^{EC}} = \frac{V_{cell}^{FC}}{V_{cell}^{EC}} \tag{15}$$

where V_{cell}^{FC} is the operating voltage in FC mode and V_{cell}^{EC} is the operating voltage in EC mode. In FC mode, RFC generally operates at 0.5–0.8 V. In EC mode, RFC operates at lower voltage for RSOFC and higher voltage for RPEMFC and RAFC as Table 3.6 shows. Assuming RFC operating at 0.7 V in FC mode, η_{cyc} of RPEMFC, RAFC, and RSOFC are 39% (@1.8 V for PEMEC mode), 35% (@2.0 V for AEC mode) and 54% (@1.3 V for AEC mode), respectively. Consequently, RSOFC reveals a distinct advantage in round-trip efficiency over low-temperature RFCs. However, a round-trip efficiency of 50%–60% is still not comparable with other energy storage technologies. Typically, SOEC operates at a voltage slightly higher than TNV (1.28 V for H_2O electrolysis). To enhance round-trip efficiency, RSOFC needs to operate at a lower voltage. Low operating voltage in EC mode decreases the electrochemical performance and leads to RSOFC turning to endothermic operating mode, which brings new challenges in cell performance and thermal management. Barnett's group from Northwestern University, USA first proposed the fuel storage cycles between H_2O-CO_2-rich feedstock gas and CH_4-rich fuel gas [77]. Using CH_4 as the storage fuel, the TNV can reduce from 1.28 V (H_2O electrolysis) to 1.04 V (H_2O/CO_2 electrolysis to CH_4). The schematic for this energy storage system combining power-to-CH_4 and direct CH_4 SOFC is shown in Fig. 3.14. According to Barnett's evaluation, an area-specific resistance of lower than 0.2 Ω cm^{-2} can realize RSOFC offering a sufficient current density (>0.5 A cm^{-2}) at a polarization voltage below 0.1 V [77]. In this case, the operating voltage in FC mode is allowed to increase to >0.9 V and that in EC mode is allowed to

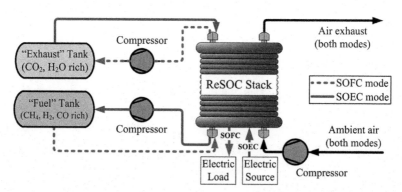

FIGURE 3.14 Schematic of RSOFC-based PtG storage with fuel cycles between CO_2/H_2O and CH_4. *(Modified from Ref. [98] Copyright 2014 Elsevier Ltd.)*

decrease to <1.1 V, hence, a round-trip efficiency can further rise to $>82\%$. Jensen et al. [97] combined this novel energy storage system with subsurface storage of compressed CO_2 and CH_4. This system enables to realize large-scale energy storage with an overall round-trip efficiency of $>70\%$ (including BOP components) and an estimated storage cost of ~ 0.03 \$ kWh^{-1}, comparable to pumped hydro storage and much more cost-efficient than other advanced energy storage technologies.

3.5   Summary

In this chapter, we introduced the category, working principles, efficiency, and other main features of different fuel cells and electrolyzers, and further discussed their contributions and compatibility when integrated in DHS. Electrolyzer is the reversed process of fuel cell. Currently, AFC, PEMFC and SOFC are the main fuel cells under development or in commercialization. Correspondingly, AEC, PEMEC and SOEC are the main electrolyzers in use or under development. To realize a more flexible conversion between electricity and fuels, a RFC is suitable energy device, which combine fuel cell and electrolyzer in one reversible reactor. Reversible alkaline fuel cell (RAEC, AFC + AEC), RPEMFC (PEMFC + PEMEC) and reversible solid oxide fuel cell (RSOEC, SOFC + SOEC) are the main RFCs under development. Hydrogen is the most common fuel used in fuel cells or generated from electrolyzers. However, the thermodynamic efficiency of H_2-fueled fuel cells was limited (83%), and the round-trip efficiency of RFC is even lower. Some fuels have shown a thermodynamic efficiency of $\sim 100\%$ such as methane, methanol, carbon, and ammonia. Among these fuels, methane is promising because of the existing natural gas infrastructures. Most of power-to-gas plants select to produce methane as the end product. A novel RSOFC-based energy storage system with fuel cycles between CO_2/H_2O and CH_4 was also proposed, which enables to realize large-scale energy storage with an overall round-trip efficiency of $>70\%$ (including BOP components) and an estimated storage cost of ~ 0.03 \$ kWh^{-1}, comparable to pumped hydro storage and much more cost-efficient than other advanced energy storage technologies.

References

[1] GE Power. The need for distributed energy solutions. https://www.ge.com/power/hybrid/distributed-energy-solutions [Accessed 01 April 2019].

[2] S. Giddey, S.P.S. Badwal, A. Kulkarni, C. Munnings, A comprehensive review of direct carbon fuel cell technology, Prog Energ Combust Sci 38 (2012) 360–399.

[3] Q. Li, J.O. Jensen, R.F. Savinell, N.J. Bjerrum, High temperature proton exchange membranes based on polybenzimidazoles for fuel cells, Prog Polym Sci 34 (2009) 449–477.

[4] L. Guo, J.M. Calo, E. DiCocco, E.J. Bain, Development of a low temperature, molten hydroxide direct carbon fuel cell, Energ Fuel 27 (2013) 1712–1719.

[5] J.C. Ganley, An intermediate-temperature direct ammonia fuel cell with a molten alkaline hydroxide electrolyte, J Power Sources 178 (2008) 44–47.

[6] C. Duan, R.J. Kee, H. Zhu, C. Karakaya, Y. Chen, S. Ricote, et al. Highly durable, coking and sulfur tolerant, fuel-flexible protonic ceramic fuel cells, Nature 557 (2018) 217–222.

[7] Department of Energy. Comparison of fuel cell technologies. https://www.energy.gov/eere/fuelcells/comparison-fuel-cell-technologies [Accessed 03 April 2019].

[8] M. Yoda, S. Inoue, Y. Takuwa, K. Yasuhara, M. Suzuki, Development and commercialization of new residential SOFC CHP system, ECS Trans 78 (2017) 125–132.

[9] Tokyo Gas. Development of the new model of a residential fuel cell, "ENE-FARM". https://www.tokyo-gas.co.jp/techno/english/category1/3_index_detail.html [Accessed 11 April 2019].

[10] Hydrogeit. Japan: Forcing the SOFC technology. 2018. https://www.h2-international.com/2018/03/06/japan-forcing-the-sofc-technology/ [Accessed 11 April 2019].

[11] T. Yoshida, K. Kojima, Toyota MIRAI fuel cell vehicle and progress toward a future hydrogen society, Electrochem Soc Interface 24 (2015) 45–49.

[12] K. Kordesch, V. Hacker, J. Gsellmann, M. Cifrain, G. Faleschini, P. Enzinger, et al. Alkaline fuel cells applications, J Power Sources 86 (2000) 162–165.

[13] Z.F. Pan, L. An, T.S. Zhao, Z.K. Tang, Advances and challenges in alkaline anion exchange membrane fuel cells, Prog Energ Combust Sci 66 (2018) 141–175.

[14] F. Bidault, D.J.L. Brett, P.H. Middleton, N.P. Brandon, Review of gas diffusion cathodes for alkaline fuel cells, J Power Sources 187 (2009) 39–48.

[15] AFC Energy reaches milestone 200 kW output from alkaline fuel cell system in Germany, 2016. http://www.renewableenergyfocus.com/view/43731/afc-energy-reaches-milestone-200-kw-output-from-alkaline-fuel-cell-system-in-germany/ [Accessed 11 April 2019].

[16] AFC Energy powers up world's first hydrogen electric vehicle charger, 2019. https://www.afcenergy.com/wp-content/uploads/2019/01/EV-Charge-PR-.pdf [Accessed 11 April 2019].

[17] AFC Energy PLC. AFC energy announces high-power density alkaline fuel cell. http://tools.euroland.com/tools/PressReleases/GetPressRelease/?ID=3592689&lang=en-GB&companycode=services [Accessed 31 May 2019].

[18] ALKAMMONIA. https://www.afcenergy.com/projects/eu-funded-projects/alkammonia/ [Accessed 11 April 2019].

[19] Y. Shi, Y. Luo, W. Li, M. Ni, N. Cai, High temperature electrolysis for hydrogen or syngas production from nuclear or renewable energy, Handbook of clean energy systems, John Wiley & Sons, Ltd., New York, (2015).

[20] W. Li, Y. Shi, Y. Luo, Y. Wang, N. Cai, Carbon deposition on patterned nickel/yttria stabilized zirconia electrodes for solid oxide fuel cell/solid oxide electrolysis cell modes, J Power Sources 276 (2015) 26–31.

[21] B. Hua, N. Yan, M. Li, Y. Sun, Y. Zhang, J. Li, et al. Anode-engineered protonic ceramic fuel cell with excellent performance and fuel compatibility, Adv Mater 28 (2016) 8922–8926.

[22] B. Hua, N. Yan, M. Li, Y. Zhang, Y. Sun, J. Li, et al. Novel layered solid oxide fuel cells with multiple-twinned $Ni_{0.8}Co_{0.2}$ nanoparticles: the key to thermally independent CO_2 utilization and power-chemical cogeneration, Energ Environ Sci 9 (2016) 207–215.

[23] T. Wan, A. Zhu, Y. Guo, C. Wang, S. Huang, H. Chen, et al. Co-generation of electricity and syngas on proton-conducting solid oxide fuel cell with a perovskite layer as a precursor of a highly efficient reforming catalyst, J Power Sources 348 (2017) 9–15.

[24] W. Li, H. Wang, Y. Shi, N. Cai, Performance and methane production characteristics of H_2O-CO_2 co-electrolysis in solid oxide electrolysis cells, Int J Hydrogen Energ 38 (2013) 11104–11109.

[25] M. Gong, X. Liu, J. Trembly, C. Johnson, Sulfur-tolerant anode materials for solid oxide fuel cell application, J Power Sources 168 (2007) 289–298.

[26] K. Huang, S.C. Singhal, Cathode-supported tubular solid oxide fuel cell technology: a critical review, J Power Sources 237 (2013) 84–97.

[27] H. Zhu, R.J. Kee, The influence of current collection on the performance of tubular anode-supported SOFC cells, J Power Sources 169 (2007) 315–326.

[28] V. Lawlor, S. Griesser, G. Buchinger, A.G. Olabi, S. Cordiner, D. Meissner, Review of the micro-tubular solid oxide fuel cell, J Power Sources 193 (2009) 387–399.

[29] Japanese group unveils SOFC Ene-Farm residential cogen unit. Fuel Cells Bulletin. 2012;2012:4.

[30] Kyocera. KYOCERA Develops Industry's First 3-Kilowatt Solid-Oxide Fuel Cell for In-stitutional Cogeneration. 2017, accessed to Apr 11, 2019. <https://global.kyocera.com/news/2017/0702_bnfo.html>.

[31] Maruta A. Japan's ENE-FARM programme. 2016, accessed to Apr 10, 2019. <https://www.energyagency.at/fileadmin/dam/pdf/veranstaltungen/Brennstoffzellenworkshop_Oktober/Maruta.pdf>.

[32] Nissan announces development of the world's first SOFC-powered vehicle system that runs on bio-ethanol electric power. 2016, accessed to Apr 11, 2019. <https://nissannews.com/es/nis-san/lac/releases/ nissan-announces-development-of-the-world-s-first-sofc-powered-vehicle-system-that-runs-on-bio-ethanol-electric-power?mode=print>.

[33] Y. Luo, Y. Shi, S. Liao, C. Chen, Y. Zhan, C. Au, et al. Coupling ammonia catalytic decompo-sition and electrochemical oxidation for solid oxide fuel cells: a model based on elementary reaction kinetics, J Power Sources 423 (2019) 125–136.

[34] M. Ryan, Methanol and fuel cells, Fuel Cell Today [Analyst views] (2012).

[35] H.M.A. Hunter, J.W. Makepeace, T.J. Wood, O.S. Mylius, M.G. Kibble, J.B. Nutter, et al. Demonstrating hydrogen production from ammonia using lithium imide—powering a small proton exchange membrane fuel cell, J Power Sources 329 (2016) 138–147.

[36] J. Cha, Y.S. Jo, H. Jeong, J. Han, S.W. Nam, K.H. Song, et al. Ammonia as an efficient CO_x-free hydrogen carrier: fundamentals and feasibility analyses for fuel cell applications, Appl Energ 224 (2018) 194–204.

[37] F. Zhou, S.J. Andreasen, S.K. Kær, J.O. Park, Experimental investigation of carbon monox-ide poisoning effect on a PBI/H_3PO_4 high temperature polymer electrolyte membrane fuel cell: influence of anode humidification and carbon dioxide, Int J Hydrogen Energ 40 (2015) 14932–14941.

[38] U.B. Demirci, Direct liquid-feed fuel cells: thermodynamic and environmental concerns, J Power Sources 169 (2007) 239–246.

[39] B.C. Ong, S.K. Kamarudin, S. Basri, Direct liquid fuel cells: a review, Int J Hydrogen Energ 42 (2017) 10142–10157.

[40] Y. Oh, S. Kim, D. Peck, D. Jung, Y. Shul, Effects of membranes thickness on performance of DMFCs under freeze-thaw cycles, Int J Hydrogen Energ 39 (2014) 15760–15765.

[41] P. Kanninen, M. Borghei, O. Sorsa, E. Pohjalainen, E.I. Kauppinen, V. Ruiz, et al. Highly ef-ficient cathode catalyst layer based on nitrogen-doped carbon nanotubes for the alkaline direct methanol fuel cell, Appl Catal B: Environ 156–157 (2014) 341–349.

[42] Panasonic, unveils high-power, durable DMFC, Fuel Cells Bull 2010 (2010) 6–7.

[43] Maryland leases GM FCV. Fuel Cells Bulletin. 2004;2004:8.

[44] Toshiba launches Dynario power source for mobile devices, but only in Japan. Fuel Cells Bul-letin. 2009:6.

[45] SFC unveils portable Jenny 1200 generator for defence market. Fuel Cells Bulletin. 2013:7-8.

[46] SFC German order to investigate DMFC use in military vehicles. Fuel Cells Bulletin. 2015;2015:4.

[47] M. Müller, N. Kimiaie, A. Glüsen, Direct methanol fuel cell systems for backup power—influence of the standby procedure on the lifetime, Int J Hydrogen Energ 39 (2014) 21739–21745.

[48] Yorkshire ambulances trialing onboard power using DMFC units. Fuel Cells Bulletin. 2011:5.

[49] J.H. Hales, C. Kallesøe, T. Lund-Olesen, A.C. Johansson, H.C. Fanøe, Y. Yu, et al. Micro fuel cells power the hearing aids of the future, Fuel Cells Bull 2012 (2012) 12–16.

[50] C.M. Miesse, W.S. Jung, K. Jeong, J.K. Lee, J. Lee, J. Han, et al. Direct formic acid fuel cell portable power system for the operation of a laptop computer, J Power Sources 162 (2006) 532–540.

[51] Ling GP. Thai Students Beat Own Record, Achieve Highest Mileage at Shell Eco-marathon Asia. 2012, accessed to Apr 20, 2019. <https://www.nationalgeographic.com/environment/great-energy-challenge/2012/thai-students-beat-own-record-achieve-highest-mileage-at-shell-eco-marathon-asia/>.

[52] Neah, Silent Falcon partner to integrate fuel cells into UAVs. Fuel Cells Bulletin. 2014:6.

[53] C. Zamfirescu, I. Dincer, Ammonia as a green fuel and hydrogen source for vehicular applications, Fuel Process Technol 90 (2009) 729–737.

[54] Agency For Toxic Substances (ATSDR). Toxic FAQ Sheet for Ammonia. 2004. p.

[55] S. Gottesfeld, The direct ammonia fuel cell and a common pattern of electrocatalytic processes, J Electrochem Soc 165 (2018) J3405–J3412.

[56] W. Zhou, Y. Jiao, S. Li, Z. Shao, Anodes for carbon-fueled solid oxide fuel cells, ChemElectroChem 3 (2016) 193–203.

[57] T. Cao, K. Huang, Y. Shi, N. Cai, Recent advances in high-temperature carbon-air fuel cells, Energ Environ Sci 1 (2017) 46–49.

[58] L. Deleebeeck, K.K. Hansen, Hybrid direct carbon fuel cells and their reaction mechanisms-a review, J Solid State Electr 18 (2014) 861–882.

[59] Jacques WW. Method of converting potential energy of carbon into electrical energy. U. S. Patent Office. Patent No. 555, 511, 1896.

[60] Cooper JF. Reactions of the carbon anode in molten carbonate electrolyte. Direct Carbon Fuel Cell Workshop, NETL, Pittsburgh, PA, July. 2003, 30. <http://w.hceco.com/dcfcws.pdf>.

[61] T.M. Gür, Direct electrochemical conversion of carbon to electrical energy in a high temperature fuel cell, J Electrochem Soc 139 (1992) L95.

[62] T.M. Gür, M. Homel, A.V. Virkar, High performance solid oxide fuel cell operating on dry gasified coal, J Power Sources 195 (2010) 1085–1090.

[63] C. Li, Y. Shi, N. Cai, Performance improvement of direct carbon fuel cell by introducing catalytic gasification process, J Power Sources 195 (2010) 4660–4666.

[64] X. Yu, Y. Shi, H. Wang, N. Cai, C. Li, A.F. Ghoniem, Using potassium catalytic gasification to improve the performance of solid oxide direct carbon fuel cells: experimental characterization and elementary reaction modeling, J Power Sources 252 (2014) 130–137.

[65] B. Heydron, S. Crouch-Baker, Direct carbon conversion: progressions of power, IOP Publishing, New York, (2006).

[66] Balachov II, Hornbostel MD, Lipilin AS. Direct coal fuel cells (DCFC): clean electricity from coal and carbon based fuels. The Carbon Fuel Cell Seminar. Palm Springs, CA, US2005. p.

[67] Wolk R, Lux S, Gelber S, Holcom FH. Direct carbon fuel cells: converting waste to electricity (report - U. S. Army corps of engineers). ERDC/CERL TR-07-32, 2007. <https://apps.dtic.mil/dtic/tr/fulltext/u2/a482934.pdf>.

[68] Gür TM. Direct carbon fuel cell with molten anode, U.S. patent office, patent application no US2006/0234098, 2006.

[69] T. Tao, L. Bateman, J. Bentley, M. Slaney, Liquid tin anode solid oxide fuel cell for direct carbonaceous fuel conversion, ECS Trans 5 (2007) 463–472.

[70] A. Jayakumar, J.M. Vohs, R.J. Gorte, Molten-metal electrodes for solid oxide fuel cells, Ind Eng Chem Res 49 (2010) 10237–10241.

[71] M. Horiuchi, S. Suganuma, M. Watanabe, Electrochemical power generation directly from combustion flame of gases, liquids, and solids, J Electrochem Soc. 151 (2004) A1402.

[72] M. Vogler, D. Barzan, H. Kronemayer, C. Schulz, M. Horiuchi, S. Suganuma, et al. Direct-flame solid-oxide fuel cell (DFFC): a thermally self-sustained, air self-breathing, hydrocarbon-operated SOFC system in a simple, no-chamber setup, ECS Trans 7 (2007) 555–564.

[73] Y. Wang, H. Zeng, T. Cao, Y. Shi, N. Cai, X. Ye, et al. Start-up and operation characteristics of a flame fuel cell unit, Appl Energ 178 (2016) 415–421.

[74] Y. Shi, N. Cai, T. Cao, J. Zhang, High-temperature electrochemical energy conversion and storage: fundamentals and applications, CRC Press, Boca Raton, (2017).

[75] S.H. Jensen, C. Graves, M. Mogensen, C. Wendel, R. Braun, G. Hughes, et al. Large-scale electricity storage utilizing reversible solid oxide cells combined with underground storage of CO_2 and CH_4, Energ Environ Sci 8 (2015) 2471–2479.

[76] S. Becker, B.A. Frew, G.B. Andresen, T. Zeyer, S. Schramm, M. Greiner, et al. Features of a fully renewable US electricity system: optimized mixes of wind and solar PV and transmission grid extensions, Energy 72 (2014) 443–458.

[77] M. Beaudin, H. Zareipour, A. Schellenberglabe, W. Rosehart, Energy storage for mitigating the variability of renewable electricity sources: an updated review, Energy Sustain Dev 14 (2010) 302–314.

[78] EG&G Technical Services. Fuel Cell Handbook. 7th ed: Contract No. DE-AM26-99FT40575; 2005.

[79] O. Schmidt, A. Gambhir, I. Staffell, A. Hawkes, J. Nelson, S. Few, Future cost and performance of water electrolysis: an expert elicitation study, Int J Hydrogen Energ 42 (2017) 30470–30492.

[80] A. Ursua, L.M. Gandia, P. Sanchis, Hydrogen production from water electrolysis: current status and future trends, IEEE 100 (2012) 410–426.

[81] C. Graves, S.D. Ebbesen, M. Mogensen, K.S. Lackner, Sustainable hydrocarbon fuels by recycling CO_2 and H_2O with renewable or nuclear energy, Renew Sust Energ Rev 15 (2011) 1–23.

[82] K.E. Ayers, E.B. Anderson, C. Capuano, B. Carter, L. Dalton, G. Hanlon, et al. Research advances towards low cost, high efficiency PEM electrolysis, ECS Trans 33 (2010) 3–15.

[83] Haldor Topsoe. Produce your own carbon monoxide. Accessed to May 28, 2019. <https://www.topsoe.com/processes/carbon-monoxide/site-carbon-monoxide>.

[84] Y. Luo, X. Wu, Y. Shi, A.F. Ghoniem, N. Cai, Exergy analysis of an integrated solid oxide electrolysis cell-methanation reactor for renewable energy storage, Appl Energ 215 (2018) 371–383.

[85] B. Paul, J. Andrews, PEM unitised reversible/regenerative hydrogen fuel cell systems: state of the art and technical challenges, Renew Sustain Energy Rev 79 (2017) 585–599.

[86] Melaina MW, Antonia O, Penev M. Blending Hydrogen into Natural Gas Pipeline Networks: A Review of Key Issues.: NREL; 2013. p.

[87] G. Gahleitner, Hydrogen from renewable electricity: an international review of power-to-gas pilot plants for stationary applications, Int J Hydrogen Energ 38 (2013) 2039–2061.

[88] M. Bailera, P. Lisbona, L.M. Romeo, S. Espatolero, Power to gas projects review: lab, pilot and demo plants for storing renewable energy and CO_2, Renew Sustain Energy Rev 69 (2017) 292–312.

[89] Y. Luo, Y. Shi, N. Cai, Power-to-gas energy storage by reversible solid oxide cell for distributed renewable power systems, J Energ Eng 144 (2018) 4017079.

[90] ZSW. P2G®-A development 'made by ZSW'. Accessed to Aug 30, 2016. <https://www.zsw-bw.de/en/research/renewable-fuels/topics/power-to-gas.html>. 2016. p.

[91] P. Schmidt, V. Batteiger, A. Roth, W. Weindorf, T. Raksha, Power-to-liquids as renewable fuel option for aviation: a review, Chem-Ing-Tech 90 (2018) 127–140.

[92] Sunfire. First commercial plant for the production of blue crude planned in Norway. https://www.sunfire.de/en/company/news/detail/first-commercial-plant-for-the-production-of-blue-crude-planned-in-norway [Accessed 31 May 2019].

[93] Y. Zhang, C. Wang, N. Wan, Z. Mao, Deposited RuO_2-IrO_2/Pt electrocatalyst for the regenerative fuel cell, Int J Hydrogen Energ 32 (2007) 400–404.

[94] M. Chen, L. Wang, H. Yang, S. Zhao, H. Xu, G. Wu, Nanocarbon/oxide composite catalysts for bifunctional oxygen reduction and evolution in reversible alkaline fuel cells: a mini review, J Power Sources 375 (2018) 277–290.

[95] Y. Wang, D.Y.C. Leung, J. Xuan, H. Wang, A review on unitized regenerative fuel cell technologies, part-A: unitized regenerative proton exchange membrane fuel cells, Renew Sustain Energy Rev 65 (2016) 961–977.

[96] S.S. Bhogilla, H. Ito, A. Kato, A. Nakano, Research and development of a laboratory scale Totalized Hydrogen Energy Utilization System, Int J Hydrogen Energ 41 (2016) 1224–1236.

[97] S.H. Jensen, C. Graves, M. Mogensen, C. Wendel, R. Braun, G. Hughes, et al. Large-scale electricity storage utilizing reversible solid oxide cells combined with underground storage of CO_2 and CH_4, Energ Environ Sci 8 (2015) 2471–2479.

[98] C.H. Wendel, P. Kazempoor, R.J. Braun, Novel electrical energy storage system based on reversible solid oxide cells: system design and operating conditions, J Power Sources 276 (2015) 133–144.

Chapter 4

High-efficiency hybrid fuel cell systems for vehicles and micro-CHPs

Yu Luo[a], Yixiang Shi[b] and Ningsheng Cai[b]

[a]National Engineering Research Center of Chemical Fertilizer Catalyst (NERC-CFC), Fuzhou University, Fujian, China; [b]Department of Energy and Power Engineering, Tsinghua University, Beijing, China

4.1 Introduction

The commercialization of fuel cells has been implemented by a number of countries and regions, involving Japan, US, Europe, China, etc. Because of no Carnot efficiency limitation, fuel cells can still remain operating with a high efficiency even at a small scale, which offers a distinct advantage over conventional heat engines. Particularly, a fuel cell with balance-of-plant (BOP) can reach an overall efficiency of ~90% when applied for combined heat and power (CHP) or combined cold, heat and power (CCHP) systems. Besides, direct fuel-to-electricity lowers the emissions of pollutant (SO_2, NOx, particulate matters, etc.) and CO_2, reduce infrastructure space and operating noisy, as well as has the ability of fast start-up and response. As a consequence, fuel cells can meet many aspects of requirement for DHS, showing great application potential in DHS. Fuel cells are applicable for stationary power generation, backup/uninterruptible power, portable power, distributed generation (such as micro-CHP or CCHP, remote area power supplies, etc.), transportation (including cars, buses, trucks, locomotives, submarines, wheel chairs, auto-bicycles, delivery vans, armoured vehicles, and small transporters at airports, shipyards, and railway stations), specialty vehicles, etc. [1,2]. Fig. 4.1 shows a typical power-efficiency curve for a fuel cell system (using PEMFC). When the load is low, the system efficiency is also quite low (~20%) in order to maintain the air supply (powering the air compressor or blower), stack cooling and inlet gas humidification. The maximum system efficiency of 60% can be achieved at a power rate of ~50%. As the power rate further increases, the system efficiency drops again due to too high polarization losses.

In this chapter, we offer some case studies on fuel cell systems applied for fuel cell vehicles and micro-CHP or CCHP.

Hybrid Systems and Multi-energy Networks for the Future Energy Internet.
http://dx.doi.org/10.1016/B978-0-12-819184-2.00004-3

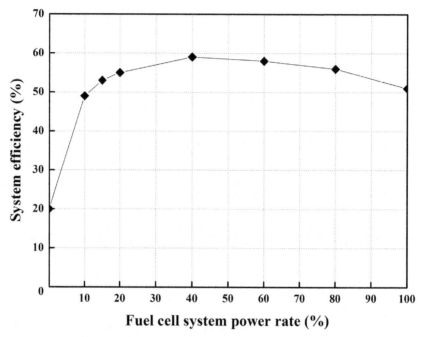

FIGURE 4.1 Typical efficiency curve of a fuel cell system [3].

4.2 Hybrid fuel cell/battery vehicle systems

Various fuel cell vehicles have been commercialized for different transportation purposes. Typical applications include cars, buses, submarines, trains, boats, forklifts, motorcycles, and material processing vehicles. Most applications se-lect proton exchange membrane fuel cells (PEMFC) because of its relatively high maturity, high performance and faster response. Typical PEMFC-based electric vehicle system is shown in Fig. 4.2. When hydrogen is used as the fuel, hydrogen directly fuels PEMFC stack to generate most of the electricity for powering the motor. A battery pack is integrated to realize fast response to transient speed-up or braking. Other fuels, such as methanol, ethanol, gasoline, etc., are also alternative fuels for fuel cell vehicles, while extra fuel process-ing subsystems like reformer or partial oxidation reactor (POX) and water gas shift (WGS) reactor are needed. Solid oxide fuel cells (SOFC) prefer a higher operating temperature, typically at 600–1000°C. This makes SOFC less flexible in start-stop and power regulation, as well as more complex BOP. However, a car typically requires a 100 kW-level capacity. In such a scale, SOFC takes the advantage in efficiency. Furthermore, SOFC is not sensitive to CO, hence, requires no CO removal processes. Nissan developed the first SOFC-powered vehicle system over the world in 2016 [4]. This system is fueled by bio-etha-nol and is able to offer the cruising ranges as high as gasoline-powered cars

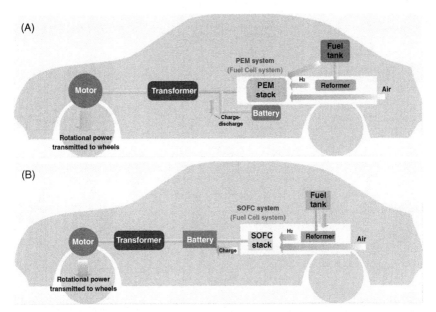

FIGURE 4.2   Schematics of PEMFC vehicle system (A) and SOFC vehicle system (B) [4].

(>600 km). A distinct feature of this SOFC-based fuel cell vehicle is that SOFC stack does not power the motor directly but charges the battery pack. Thus, the users can obtain the driving experience similar to pure electric vehicle.

4.2.1   PEMFC-based fuel cell vehicle systems

Despite hydrogen is widely believed to be the ultimate aim for fuel cell cars, there is still a debate on how to utilize hydrogen. The onboard compressed hydrogen storage can simplify the fuel cell vehicle system while requiring higher costs for refueling infrastructures and hydrogen storage. Another solution is to store hydrogen in H-containing fuels such as methanol, ethanol, methane, and gasoline, and convert them into hydrogen onboard the vehicle through a series of fuel processing. Ogden et al. [5] from Princeton University carried out an early and comprehensive study to compare hybrid fuel cell/battery vehicle systems fueled with diverse fuels, including hydrogen, methanol, and gasoline in the electrochemical performance, fuel economy, cost, infrastructure requirements, and lifecycle cost of transportation.

4.2.1.1   System diagram

The system diagrams for these three fuels-fueled fuel cell vehicles are shown in Fig. 4.3A–C. These systems aimed to power a PNGV (the Partnership for a New Generation of Vehicles)-type midsize automotive car with reduced weight, rolling resistance, and aerodynamic drag [5]. Ogden built a numerical simulation

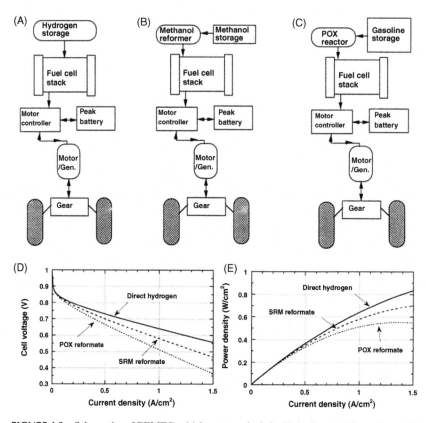

FIGURE 4.3 Schematics of PEMFC vehicle systems fueled with hydrogen (A), methanol (B) and gasoline (C), and their performance curves (D and E). *(Modified from Ref. [5]. Copyright 1999 Elsevier Ltd.)*

model to simulate these three systems including dynamic PEMFC model, steam reformer, POX reactor, WGS reactor, and selective catalytic oxidation (SCO) reactor. This system model considered a variety of input parameters including the driving schedule (urban driving schedule and highway driving schedule), vehicle parameters (basic weight), fuel cell system parameters (polarization curves at various inlet H_2 content, fuel utilization ratio, capacity, and weight), peak power battery characteristics (charging-discharging characteristics, capacity, and weight) as well as fuel processing parameters (conversion efficiency, response time, weight), etc. When the fuel cell is unable to power the motor alone, the power battery can replenish extra needed power. This system will allocate the power demand to the fuel cell and battery according to the principle that the battery should be maintained near an ideal state of charge (SOC). This model calculated the consumed fuel according to the efficiency of fuel processing subsystem and polarization curve of PEMFC stack. The fuel economy was obtained by calculating the total fuel consumption over a period. These three hybrid systems were assumed to have 50 kg fuel tank, where can store 3.75 kg H_2 (700 bar) or 13 gal methanol or gasoline. The PEMFC stack operated at 3 atm with an excess air coefficient of 2. For hydrogen-fueled system, pure H_2 was fed into the anode of PEMFC stack. But methanol or gasoline-fueled system, the inlet H_2 content is up to the output of fuel processing subsystem. The approximate inlet H_2 content is ~75% for methanol-fueled system and ~40% for gasoline-fueled system. Fig. 4.3D gives the polarization curves of PEMFC fueled with pure hydrogen, methanol reformate and gasoline POX, and Fig. 4.3E gives the corresponding power density curves. The curves reveal that ~75% H_2-containing fuel gas can support PEMFC to generate a power density of 0.70 W cm^{-2}, 83% of pure H_2 case, while ~40% H_2-containing fuel gas can only offer 0.56 W cm^{-2}, 68% of the power density in pure H_2 case.

Methanol and gasoline are both liquid fuels at room temperature and normal pressure, hence, offering higher volumetric energy density. The system layouts for onboard liquid fuel-fueled hybrid fuel cell systems are shown in Fig. 4.4 [5]. Compared with H_2-fueled system, hydrocarbon-fueled system requires extra fuel reformer or POX reactor, WGS reactor, SCO reactor, and other BOPs like heat exchangers, burners, compressors, and turbines. Therefore, this system is much more complicated. Fig. 4.4A shows that methanol (CH$_3$OH) was mixed with water and reformed into H_2- and CO-rich fuel gas. CO contained in fuel gas was converted into CO_2 in WGS reactor and oxidized to a ppm-level volumetric fraction in an SCO reactor. A fuel gas containing ppm-level CO was tolerable for PEMFC and can be fed into anode directly. The exhausted gas from

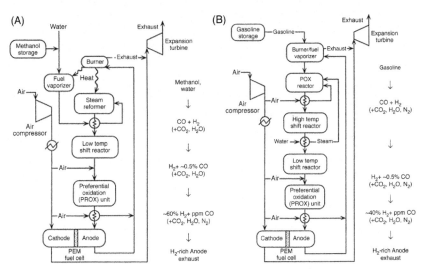

FIGURE 4.4   **System layouts for onboard methanol-fueled (A) and gasoline-fueled (B) hybrid fuel cell systems.** *(Modified from Ref. [5]. Copyright 1999 Elsevier Ltd.)*

TABLE 4.1 Technical parameters for different fuel-fed fuel cell vehicle systems [5].

Fuels	System mass (kg)	Peak power (kW) (FC/battery)	Fuel efficiency (MPGe)	Driving distance (km)
Hydrogen	1170	77.5 (34.4/43.1)	106	684
Methanol	1287	83.7 (37.0/46.7)	69	740
Gasoline	1395	89.4 (39.4/50.0)	71	1513

anode was fed into a burner to offer heat for fuel vaporizer and steam reformer. The methanol-to-H_2 efficiency for methanol reformer is equal to the ratio of Higher Heat Value (HHV) of produced H_2 in methanol reformer to that of inlet methanol, ~77.5% in value. Considering an H_2 utilization of 80%, the overall efficiency for methanol processing subsystem is equal to 62.0%. As for gasoline-fueled fuel cell vehicle, gasoline processing used a POX reactor, which is thermally self-sustainable. The heat from burning anode exhausted gas can be used for gasoline vaporization or supporting expansion turbine. In this system, the gasoline-to-H_2 efficiency is 86.7% and overall efficiency is equal to 69.4%.

The power battery pack was the spiral wound, thin film lead-acid batteries (12 Ah for each one) connected in series. The power battery pack had a current limitation of 30 A and energy density of 15 Wh kg^{-1}. The power battery weight is approximately 1.0 kg for each kW. The energy was captured when braking, and the SOC of batteries was kept at ~0.5. When SOC is higher than 0.5, the system allowed the battery pack to offer more power. When SOC is lower than 0.5, the PEMFC stack generated more power and charged the battery pack when low power demand. Ogden et al. [5] also compare the effect of the response time (1 and 5 s) of fuel processing subsystem on the whole fuel cell vehicle system. A response time of 1 s allows PEMFC stack to match well with the power demands, while a response time of 5 s reveals an apparent lag of PEMFC stack in load following. The lag means that the power battery pack is more likely to be used and the energy offered from battery also increases. According to Ogden's calculation, 40%–50% of energy could be directly supplied by the power battery in urban application. A surge tank storing reformed H_2-rich fuel gas could be an feasible solution to shorten the time delay caused by fuel processing subsystem.

4.2.1.2 System weight and fuel efficiency

According to Ogden's models and assumptions, the system mass, capacity requirement of PEMFCs and power batteries, fuel economy, and driving distance are summarized in Table 4.1 [5]. These parameters were evaluated in the premise of the same power output and characteristics.

Fig. 4.5A shows the distribution of system weight for three fuel cell vehicle systems. The system mass for hydrogen, methanol, and gasoline was different due to the difference in fuel processing subsystem and required capacity of PEMFCs and power batteries. Producing H_2 onboard inevitably increases the system weight. On one hand, fuel processing subsystem and additional support structure brought extra weight. On the other hand, PEMFC stack in liquid fuel-fed fuel cell vehicle was fueled by the fuel gas with lower H_2 content, which reduced the performance and efficiency of PEMFC stack as Fig. 4.2D and E reveals. Hence, these fuel cell vehicle system requires larger PEMFC stack to meet the same power demand. As a result, the system mass of gasoline-fueled fuel cell system is 1395 kg, the heaviest among these three vehicle systems, 108 kg (10%) heavier than methanol-fueled one and 225 kg (19%) heavier than hydrogen-fueled one.

In these three fuel cell vehicle systems, the peak power of PEMFC stack and power battery pack is closed. Hydrogen-fueled system requires the smallest

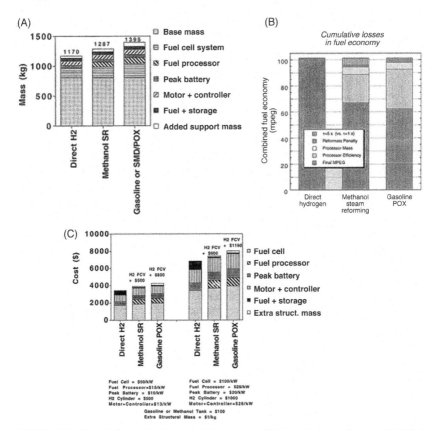

FIGURE 4.5 **Distributions of vehicle weight, energy efficiency losses and costs.** *(Modified from Ref. [5]. Copyright 1999 Elsevier Ltd.)*

peak power, only 34.4 kW for PEMFC and 43.1 kW for power battery because of lighter system. Actually, power demands for urban driving and high-way driving are generally less than 30 kW. The power demands could increase to higher than 50 kW in the extreme conditions such as hill climbing. The power battery was used to offer the extra power demands in these special cases. In the hydrogen-fueled system, the peak power for the power battery (43.1 kW) was used to recover the energy during emergent braking. But for the liquid fuel-fed systems, the power battery could be frequently used to deal with the time delay of fuel processing subsystem. Therefore, the peak power rose to 46.7 kW for methanol-fueled fuel cell vehicle and 50.0 kW for gasoline-fueled one.

According to Ogden's model [5], the hydrogen-fueled fuel cell system for PNGV-type midsize automobiles had a fuel efficiency of 106 MPGe (miles per gallon gasoline equivalent), while the methanol-fueled and gasoline-fueled ones were at least 1/3 lower in the fuel efficiency, that is, 69 and 71 MPGe, respectively. The reasons for the losses of fuel efficiency were analyzed as Fig. 4.5B shows. The primary reason is a decrease (17%–32%) in the electrochemical performance of PEMFC due to lower H_2 content, leading to the fuel efficiency of the methanol-fueled system dropping by ~21 MPGe and that of the gasoline-fueled one dropping by ~30 MPGe. An increase of 10%–20% in system weight due to onboard fuel processing subsystem also caused a decrease of ~5 MPGe in fuel efficiency. For methanol-fueled fuel cell vehicle system, the slow response enhanced the power supply from the power battery, which brought extra round-trip efficiency losses (a decrease of ~6 MPGe). The efficiency losses (15%–24%) during hydrogen production from methanol or gasoline also caused a decrease of 1–2 MPGe in fuel efficiency. According to the fuel efficiency, the driving distance can be calculated according to the fuel storage tank. Despite almost 50% higher fuel efficiency of hydrogen-fueled fuel cell vehicle, the low volumetric density of compressed H_2 (at 700 bar) led to the shortest driving distance (684 km) of hydrogen-fueled system among these three systems. The methanol-fueled one had an estimated driving distance of 740 km, 8% higher than hydrogen-fueled one. The driving distance of the gasoline-fueled one can even reach over 1500 km, 2.2 folds of the hydrogen-fueled one. As mentioned in Ref. [5], the target driving distance of PNGV is ~610 km. All the considered fuel cell vehicle systems satisfied this target. However, the electrochemical performance of the PEMFC was determined according to the experimental data, which could be lower after scaling up. Compared with current commercialized fuel cell vehicles like Mirai (69 MPGe), the estimated fuel efficiency of Ogden's study was much higher.

4.2.1.3 Hybrid ratio

The fuel efficiency is not only dependent on the type of fuel, but also determined by the installed capacity of fuel cells and power batteries. As Fig. 4.1 shows, a fuel cell system prefers to operate at a power rate of ~50% for higher system efficiency. A power battery can recover the kinetic energy from braking and allows the fuel cells to operate at their preferred operating condition, but the frequent use of power battery also leads to the extra losses due to the round-trip efficiency of the power batteries. Jeong and Soo [3] defined an indicator named hybrid ratio to represent the hybridization level of fuel cells and power batteries. The hybrid ratio can be calculated as below:

$$\text{Hybrid ratio} = \frac{P_{tot}^{max} - P_{FC}^{max}}{P_{tot}^{max}} \tag{1}$$

where P_{tot}^{max} is the maximum power output of the whole vehicle system, and P_{FC}^{max} is the maximum fuel cell power output.

Jeong et al. compare the fuel efficiency of a hydrogen-fueled fuel cell vehicle system at the hybrid ratio of 0–0.733 as Fig. 4.6 shows. Jeong's study reveals that there is an optimal hybrid ratio in a hybrid fuel cell/battery vehicle system. When the hybrid ratio is too low, the capacity of fuel cell stack is high, leading to fuel cells operating at a relatively low power rate in most of cases (unless hill climbing). This results in a lower fuel cell system efficiency. As the capacity of power battery rises, the fuel efficiency also increases because the fuel cell stack is able to operate at its preferred power rate and the power battery can also recover the kinetic energy during the braking. If the capacity of fuel cell stack is too low, the power battery has to supply the power demand more frequently, that is, a large amount of energy for vehicle driving comes from the power battery. The charge-discharge cycles lead to extra round-trip efficiency losses. As a result, the fuel efficiency at a hybrid ratio of 0.733 is even 13% lower than that using pure fuel cell system. Consequently, the optimal hybrid ratio is ~0.333 according to Jeong's calculation.

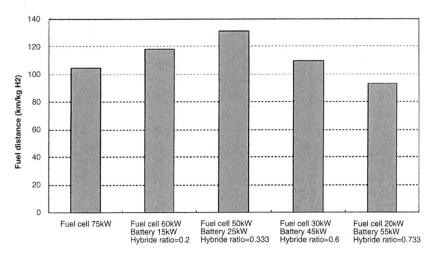

FIGURE 4.6   Fuel efficiency of a H$_2$-fueled hybrid fuel cell/battery vehicle system at various hybrid ratio. *(From Ref. [3]. Copyright 2002 Elsevier Ltd.)*

4.2.1.4 Investment cost and fuel cost

Based on these three fuel cell vehicle systems, Ogden et al. [5] also estimated the overall system investment costs as Fig. 4.5C shows. The costs for the fuel cell subsystem, fuel processing subsystem, power battery pack, motor, controllers, fuel storage, and extra structures were considered. The components evaluated two prices, that is, the high estimate one and the low estimate one. For example, the PEMFC stacks were estimated to be $50 per kW in the low estimate case and $100 per kW in the high estimate case. Gasoline or methanol tank was considered to be constant, equal to $100 for one. According to Fig. 4.5C, hydrogen-fueled fuel cell vehicle system has a total low estimate cost of $3600 and a high estimate one of $7000. The half of the total costs came from the PEMFC stacks, and the motor and controller were the second highest costs. The fuel storage and power batteries had similar costs, both accounting for 12%–14% of the total investment costs. When methanol or gasoline was used as the stored fuel, the total investment increased by $500–600 for methanol-fueled fuel cell vehicle system and by $850–1200 for gasoline-fueled one [5]. Ogden assumed that the fuel processing subsystem costs of methanol-fueled system and gasoline-fueled system are similar. The high investment costs for the gasoline-fueled fuel cell vehicle system were mainly attributed to their poor electrochemical performance of PEMFC stack at low H_2 content. To realize the same power output, the gasoline-fueled fuel cell vehicle system needed larger size of PEMFC stacks and other components, and also increased the system weight. As Ogden mentioned, the reference cost for a gasoline internal combustion engine (ICE) vehicle was approximately $39 per kW and the total costs for a typical car with 94 kW engine should be $3670 [5]. Consequently, the costs of these components used in fuel cell vehicle systems should be reduced to approach to the low estimate in Ogden's study.

To summarize, hydrogen-fueled fuel cell vehicles have simpler system structure, less system weight, higher electrochemical performance and fuel efficiency, as well as lower investment costs. However, methanol-fueled or gasoline-fueled fuel cell vehicles have longer driving distance due to high volumetric energy density of liquid fuels. Moreover, the cost for fuel and refueling infrastructure should be also estimated. According to Ogden's study, Table 4.2 summarizes the annual fuel consumption for these three types of fuel cell vehicles by assuming that driving distance were 17,600 km per year. After considering the fuel efficiency, one hydrogen-fueled fuel cell vehicle consumes the least energy (determined by HHV of fuels), that is, 13.6 GJ per year, among these three types of fuel cell vehicles. The consumed energy of the methanol-fueled and gasoline-fueled fuel cell vehicles are quite close. The former one was 20.9 GJ per year while the latter one was 20.3 GJ per year. According to current fuel price (2019), hydrogen costs ~$7 for each kilogram, methanol costs $1.26 per gallon, and gasoline costs almost twice of methanol. Therefore, the annual fuel costs for H_2, methanol, and gasoline are $677, $384, and $388, respectively. The fuel cost for hydrogen-fueled fuel cell vehicles is more than

TABLE 4.2 Fuel consumption and annual fuel cost [5].

Fuels	Hydrogen-fueled	Methanol-fueled	Gasoline-fueled
Fuel economy (MPGe)	106	69	71
Driving distance per year (km)	17,600		
Stored fuel	3.75 kg H$_2$	13 gal methanol	13 gal gasoline
Energy from the annually consumed fuel (GJ/year)	13.6	20.9	20.3
Annual fuel consumption	1130 Nm^3 = 96.7 kg	307 gal = 919 kg	155 gal = 587 L
Fuel price	$7 kg^{-1}	$1.26 gal^{-1}	$2.50 gal^{-1}
Annual fuel cost	$677	$384	$388

74% more expensive than that for methanol- or gasoline-fueled fuel cell vehicles. Assuming the same maintenance cost and lifetime, methanol-fueled fuel cell vehicles became the most economically feasible among these three fuel cell vehicles after at least 2-year use. The gasoline-fueled fuel cell vehicles are more economic than the hydrogen-fueled fuel cell vehicles after a 3- to 4-year use. Of course, this situation should be re-evaluated if carbon emission penalty is put into consideration.

4.2.2 SOFC-based fuel cell vehicle systems

According to the Nissan's hybrid SOFC/battery vehicle type (Nissan Leaf Acentan SOFC), Bessekon et al. [6] built a simulation platform for the hybrid SOFC/battery vehicle systems fueled with compressed natural gas (CNG), liquefied natural gas (LNG), and liquefied petroleum gas (LPG). These systems had a fuel tank with 30 L volume and installed an SOFC subsystem. The SOFC subsystem consists of a 12 kW SOFC unit and a POX reformer. Differently from hybrid PEMFC/battery vehicle system, hybrid SOFC/battery vehicle system uses the power battery to supply all the power loads directly. Therefore, this hybrid SOFC/battery vehicle has a similar driving experience to a pure battery-driven electric vehicle system. However, the SOFC unit is used to charge the power battery directly so that the driving distance can be significantly enhanced.

4.2.2.1 Power control and management

Fig. 4.7A shows the algorithm of the power control and management for the hybrid SOFC/battery system [7]. The motor is only driven by the power battery, and the power battery can operate in three modes: hybrid mode, discharging

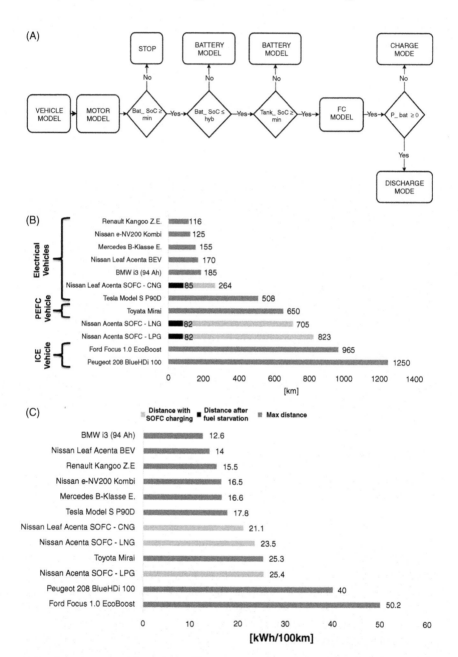

FIGURE 4.7 (A) Power control and management of the hybrid SOFC/battery vehicle system. (B) Driving distances of various vehicles. (C) Fuel efficiency of various vehicles. *(Modified from Ref. [3]. Copyright 2018 Elsevier Ltd.)*

mode, and stopping mode. The power battery has two critical SOC values for controlling, where the lower one is the minimum SOC (SOC_{min}) and the upper one is the hybrid SOC (SOC_{hyb}). When the real-time SOC is lower than SOC_{min}, the whole system has to stop operating (stopping mode) due to the shortage of electrical power. When the real-time SOC is higher than SOC_{hyb}, the power battery is full and SOFC does not need to work (discharging mode). When the real-time SOC is between SOC_{min} and SOC_{hyb}, the power battery requires SOFC to charge. Only if the fuel tank contains enough fuel, the system operates in the hybrid mode. In the hybrid mode, SOFC operates at a steady condition to offer the nominal power output, while the battery offsets the remaining power demand or stores the surplus power output.

4.2.2.2   Driving distances

Fig. 4.7B compares the driving distances of a wide range of ICE vehicles, battery-powered electrical vehicles and hybrid fuel cell/battery vehicles. For comparison, Bessekon et al. [6] uniformed the volume of fuel tanks to 30 L (closer to a standard tank size of ICE vehicles). Correspondingly, the capacity of the SOFC unit was enlarged to 12 kW. CNG, LNG, and LPG were considered as the feeding fuels in the hybrid SOFC/battery system, respectively. As estimated, the system weights using these three fuels are quite similar, that is, 1456 kg for CNG, 1459 kg for LNG, and 1460 kg for LPG. In Fig. 4.7B, the red columns denote that the driving distance without SOFC discharging (only in charging mode), and the total length of red column and green column denotes the driving distance of hybrid SOFC/battery vehicle system. Traditional ICE vehicles have the longest driving distance of all the listed vehicles, typically exceeding 900 km. For other currently commercialized vehicles, Toyota Mirai has longer driving distance (650 km) than other pure battery-powered electrical vehicles. Tesla Model S is the battery-powered electrical vehicle with the longest driving distance, as far as 508 km. Without the integration of SOFC, the Nissan Leaf Acenta SOFC can only maintain a ~80 km driving distance due to an SOC_{hyb} of 0.5. The charging of SOFC enhances the driving distance remarkably. The hybrid SOFC/battery system fueled with CNG can realize a driving distance of 264 km, 3.1 folds of the one without SOFC discharging. When LNG or LPG is fueled with this hybrid system, the driving distance has a much more significant increase by 8.6–10.0, reaching more than 700 km and even exceeding all current battery-powered electric vehicles and PEMFC vehicles.

4.2.2.3   Energy consumption and fuel efficiency

Fig. 4.7C compares the hybrid SOFC/battery vehicle system with other commercialized vehicles in fuel efficiency. The hybrid SOFC/battery vehicle system drives the motor directly through the power battery, hence, the fuel efficiency of the hybrid SOFC/battery vehicle system is the result of SOFC discharging, battery charging and battery discharging. As for the battery-power electric vehicles, their efficiency is only determined by battery discharging. The efficiency of battery-power electric vehicles is not a fuel-to-wheel efficiency, but a battery-to-wheel efficiency. The extra losses from SOFC discharging and battery charging lead to lower energy efficiency of the hybrid SOFC/battery vehicle. Therefore, it is not fair to compare the energy efficiency of battery-powered electric vehicles and fuel cell-based vehicles based on the average energy consumption per 100 km. The efficiency losses of battery charging (from charging piles) and upstream power generation (Municipal electric power from power plants or other energy sources) are not considered onboard. As for the ICE vehicles and the hybrid PEMFC/battery vehicles, their fuel efficiencies are comparable with that of the hybrid SOFC/battery vehicles. The CNG-fueled hybrid SOFC/battery vehicle system has an energy consumption of 21.1 kWh per 100 km, 17% lower than that of Toyota Mirai (hybrid PEMFC/battery vehicle system) and approximately an half of that of the ICE vehicles. When the hybrid SOFC/battery vehicle system uses LNG or LPG as the fuel, the energy consumption rises by 10%–20%, meaning a decrease in fuel efficiency.

Consequently, Bessekon's study concluded that the hybrid SOFC/battery vehicle system performs quite well in driving distance and fuel efficiency, better than the Toyota Mirai's PEMFC-based vehicle system. However, it is still too rash to conclude that hybrid SOFC/battery systems perform better than hybrid PEMFC/battery systems due to the difference in fuel type. Toyota Mirai is fueled with compressed hydrogen, which has a lower volumetric energy density than CNG, LNG, and LPG, resulting a shorter driving distance. Therefore, a comparable system simulation platform and evaluation indicator should be built to give a more comprehensive and fairer comparison among PEMFC-based fuel cell vehicles, SOFC-based fuel cell vehicles and pure battery-powered electric vehicles.

SOFC allows to directly convert CO-rich fuel gas, hence, show high compatibility with some common hydrocarbon fuels. Therefore, some existing refueling infrastructure or natural gas pipeline are available for cost saving. Based on these infrastructures, the hydrocarbon-fueled hybrid fuel cell/battery vehicles can maintain long driving distance with a high fuel efficiency and also reduce CO_2 emission, hence, potentially accelerate the updating of the ICE vehicles. From the perspective of fuel production, hydrocarbon fuels can be also produced from renewable energy sources through power-to-gas or power-to-liquid technologies. By combining these technologies, a carbon-neutral energy conversion pathway is formed to offer a sustainable energy utilization pattern.

4.2.2.4 Effect of fuel storage volume, SOFC performance and battery capacity

In a hybrid SOFC/battery vehicle system, an increase in the fuel storage volume is similar to an increase in the battery capacity. Fig. 4.8A and B shows the effect of fuel storage volume on the driving distance and fuel efficiency. Fig. 4.8A shows the driving distance rises linearly with the tank volume increasing. A tank volume of 60 L can enhance the driving distance of a hybrid CNG-fueled SOFC/battery vehicle to ~400 km, that of a LNG-fueled one to

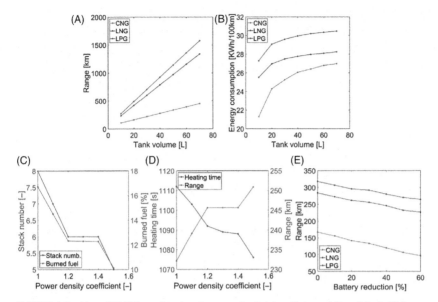

FIGURE 4.8 **(A and B) Effect of tank volume on the driving distance (A) and fuel efficiency (B). (C and D) Effect of SOFC performance on the required stack number and burned fuel (C) and on the heating time and driving distance (D). (E) Effect of battery capacity on the driving distance.** *(Modified from Ref. [3]. Copyright 2018 Elsevier Ltd.)*

1200 km and that of a LPG-fueled one to 1400 km. Fig. 4.8B reveals that the energy consumption slightly increases with tank volume increasing, denoting the fuel efficiency dropping. This is because large fuel tank volume enables SOFC to generate the nominal power for a longer time according to the system control strategy. When SOFC generate the nominal power, the energy consumption is larger.

Currently, SOFC is still developing, and SOFC performance is expected to keep rising with its rapid development. Fig. 4.8C and D evaluates the effect of SOFC performance on the hybrid SOFC/battery vehicle system. Higher SOFC performance (higher power density) allows SOFC to generate more power in a restricted volume or effective reaction area and operate more efficiently. Therefore, a rise in power density will lead to reducing the number of SOFC stacks and increasing the driving distance. Fig. 4.8C indicates that the required SOFC stack number reduces from 8 to 5 as SOFC performance rises by an half, which reveals a great potential for saving the investment costs. Meanwhile, better SOFC performance reduces the burning fuel for heating by six points of percentage and shortens the heating time by ~30 s because of the reduction of the SOFC volume (lower heat capacity). Finally, the driving distance enhances slightly by 8% with power density rising by an half, which is mainly limited by the fuel storage capacity.

In the hybrid SOFC/battery system, the power battery is vital as the direct power supplier of the powertrain. When the SOFC only generates the nominal power, the power battery is the only adjustable power supplier. Fig. 4.8E shows the effect of the battery capacity curtailment on the driving distance. A reduction by 60% in the battery capacity leads to the driving distance dropping from 170 km by 42% to 98 km when CNG is used as the fuel. When LNG and LPG are respectively used, the driving distance drops only by 20% and 17% with the battery capacity curtailing by 60%. This means that the driving distance does not decreases linearly with the battery capacity curtailment. In the premise of enough peak power output, a part of power battery can be curtailed, instead, the fuel tank can be enlarged for longer driving distance.

4.3 Fuel cell-based micro CHP or CCHP systems

Conventional centralized power plants have long lifetime and low flexibility in upgrading, leading to the "carbon lock-in effect" [8]. A distributed energy system (DES) shows a high flexibility in upgrading and scale selection, and offer a feasible pathway for balancing power supply and demand even at high renewable energy penetration. In a common sense, a micro-CHP or CCHP system is a typical DES, which is a key technology to allow higher utilization of renewable energy sources. In smart grids, micro-CHP or CCHP is served as a basic unit and connected to form a so-called Virtual Power Plant [8], a novel concept and solution to transform energy utilization pattern from centralization to distribution. Micro-CHP or CCHP can offer diverse forms of energy including heat, electricity, cold and even gas, hence, is more flexible in energy dispatch, multiform energy storage (electricity, heat or cold) and cascading utilization. Most of current micro-CHPs in use are based on the heat engines, involving Stirling engine, Organic Ranking Cycle, ICE or gas turbines cycle. Fuel cell-based micro-CHP or CCHP systems reveal competitiveness because of their high efficiency particularly at a small-scale application. It is reported by FuelCellToday that the sold fuel cell-based micro-CHP or CCHP systems have accounted for 64% of global micro-CHP or CCHP sales in 2012, exceeding conventional engine-based ones.

The micro-CHP or CCHP systems as DES are more likely to aim at the end-use customers. Therefore, micro-CHPs or CCHPs should have a timely response to the energy demand of the end-use customers. Diverse end-use customers have different energy demand over time. Taking a residential application as an example, Adam et al. [9] give a representative energy demand curves over time in the UK on a winter day as Fig. 4.9 shows. The activities of residents have a significant effect on the energy demands. The demand peak for heating, hot water, and electricity is attributed to residents waking up. The demands keep low in the daytime because residents are outside for work. In the evening, temperature drop outside and residents coming home result in a higher heating demand peak than the peak in the morning. Unlike electricity demand, heat demands including heating demand and hot water demand allow a longer time delay due to the delay of human senses. However, heat demands account for the majority of the total energy demands, about 80% or even higher.

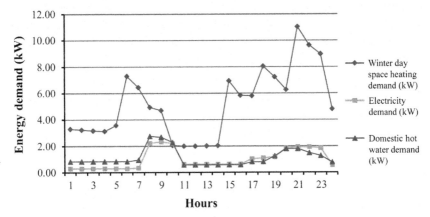

FIGURE 4.9 Residential demand for heating, electricity and hot water over time in the UK on a representative winter day. *(From Ref. [9]. Copyright 2014 Elsevier Ltd.)*

4.3.1 Basic schematic diagram of fuel cell-based micro-CHP or CCHP systems

Typically in residential houses, heat demands require a temperature of 30–90°C [9]. Conventional heating systems use gas-fired boilers or heat pumps for heat generation. Micro-CHP systems allow the integration of renewable energy sources like solar energy and geothermal sources to supply the heat. PEMFC and SOFC are all optional fuel cell for micro-CHP or CCHP. SOFC operates at a much higher temperature than PEMFC, hence, shows an advantage in both the electricity efficiency and the overall efficiency when applied for a cogeneration system.

Because of the existing natural gas pipeline, natural gas is more suitable as the fuel for micro-CHP or CCHP systems, particularly in the residential applications. Fig. 4.10 shows the typical schematic of PEMFC-based and SOFC-based micro-CHP systems [8]. Fig. 4.10A reveals a PEMFC-based micro-CHP system layout fueled with natural gas, where PEMFC stacks generate both electricity and heat with a temperature of <100°C. Natural gas is first fed into a fuel processing subsystem (including a reformer and a WGS reactor) to produce H_2-riched fuel gas with ppm-level CO content. The heat demand of natural gas reforming is supplied by an integrated burner. The fuel for the burner comes from two parts: one is the natural gas from the gas network, and the other one is the unfired fuel from the exhausted gas. Apart from the heat recovered from PEMFC stack, the heat can be also recovered from the reformer outlet and burner outlet. Owing to the temperature limitation of PEMFC, the reformed fuel gas must be cooled down to lower than 100°C. The recovered heat is connected to a heating system for room heating and hot water supply. Fig. 4.11 shows a typical schematic of a heating system. A storage tank is used in the heating system for hot water buffering and storage.

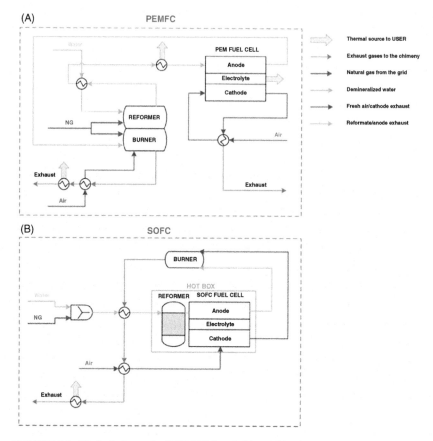

FIGURE 4.10 Typical schematic of PEMFC-based (A) and SOEC-based (B) micro-CHP systems. *(From Ref. [8]. Copyright 2015 Elsevier Ltd.)*

Fig. 4.10B demonstrates an SOFC-based micro-CHP system layout fueled by natural gas network. In a natural gas-fueled SOFC-based micro-CHP system, a simpler fuel processing subsystem is required because of two reasons. One reason is that SOFC is capable of converting CO and other hydrocarbons like CH_4 (main content of natural gas) directly. To guarantee the enough lifetime, natural gas still needs to be pretreated by desulfurization and steam reforming to avoid sulfur-poisoning and carbon deposition in SOFC anode. However, extra CO removal can be eliminated from the micro-CHP system. The other reason is that SOFC and reformer match with each other in temperature. Consequently, the reformer and SOFC can be integrated in one hot box, that is, so-called internal-reforming SOFC reactor. The operation of SOFC produces high temperature heat, which can be immediately used for supporting strongly endothermic steam reforming. This compact system configuration can significantly reduce the system complexity and offer higher efficiency due to better thermal coupling. Compared with PEMFC-based micro-CHP system, SOFC-based micro-CHP can also offer the heat with higher grade, and only needs to recover the heat from the outlet of after-burner. Fig. 4.11B shows the electrical efficiency, thermal efficiency, and overall efficiency at various load factor. The electrical demand and heat demand of a building is up to different building functions. The SOFC-based micro-CHP reveals its capability in regulating the ratio of electricity to heat. The electrical efficiency reaches the peak, ~50%, at a load factor of 30%, while the corresponding overall efficiency is only 70% at this load factor. As the load factor increases, the electrical efficiency drops slightly. The electrical efficiency is around 42% at the full load. The thermal efficiency and overall efficiency both increases significantly as the load factor rises. At the full load, the thermal efficiency can reach 38%, and the overall efficiency can reach 80%. The ratio of electricity to heat is the highest at a load factor of 30%, reaching ~2.5. The ratio reaches the lowest at the full load, only 1.1. In real operation, the minimum power output of a fuel cell system is generally no less than 20%–30% of the nominal power. Meanwhile, the variation of power output should not be too fast, or the failure of fuel cells could be caused due to drastic temperature variation. Typically, the ramp up rate of a SOFC-based micro-CHP system can reach 9 kW_e min^{-1} [9]. Thus, when the heat demand is quite high, we can equip the micro-CHP system with a separate boiler to offer extra heat demand so that the electrical demand and heat demand can be decoupled.

4.3.2 Direct flame solid oxide fuel cell for micro-CHP or CCHP systems

Direct flame fuel cell (DFFC) is a novel single-chamber fuel cell combining a fuel flame with SOFC, where anode is exposed in the fuel-rich flame region and cathode in air. DFFC has a series of distinct features including fuel flexibility, no need for sealing, simple configuration and rapid start-up [10]. Due to high-grade heat generated from fuel combustion, DFFC is quite suitable for the residential micro-CHP or CCHP. Wang et al. [11] proposed a micro-CCHP system based on direct flame SOFC (DF-SOFC) as Fig. 4.12 shows, in which

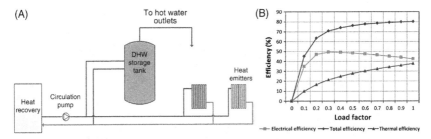

FIGURE 4.11 (A) Domestic heating system for heating supply and hot water supply; (B) Electrical efficiency, thermal efficiency and overall efficiency of an SOFC-based micro-CHP system. *(Modified from Ref. [9]. Copyright 2014 Elsevier Ltd.)*

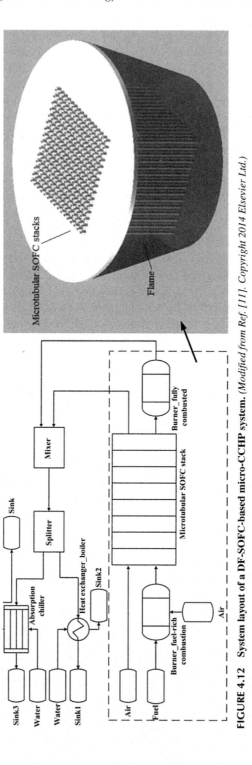

FIGURE 4.12 **System layout of a DF-SOFC-based micro-CCHP system.** *(Modified from Ref. [11]. Copyright 2014 Elsevier Ltd.)*

a boiler and a chiller were integrated for enhancing the flexibility in the generation of heat and cold. In this system, a micro-tubular DF-SOFC stack is integrated in a fuel-rich burner. The burner is also the anode chamber. Natural gas is mixed with a small amount of air and form fuel-rich flame outside the micro-tubular DF-SOFC stack (anode chamber). DF-SOFC generates DC electricity and high-grade heat ($\sim$900°C). The exhausted gas with high-grade heat is separated into two streams. One stream exchanges heat with water in a boiler to offer hot water for residents, and the other one is fed to a double-effect LiBr ater adsorption chiller for cooling water (as air conditioning).

Based on the proposed system, Wang et al. analyzed the effects of equivalence ratio and fuel utilization on the overall efficiency and the ratio of heat to electricity. Simulation results reveal that a rise from 1.4 to 2.0 in the equivalent ratio reduces the electrical efficiency of DF-SOFC by 2.5% to 26.2% but enhances the power output from 0.72 to 1.14 kW. As for the whole system, the increase of the equivalent ratio enhances the overall efficiency slightly and decreases the ratio of heat to electricity from 7.9 to 4.5. A rise in the fuel utilization in DF-SOFC stack can apparently increase both the power output and electrical efficiency. The overall efficiency also enhances with the fuel utilization in DF-SOFC stack, but increase insignificantly as the fuel utilization increases from 60% to 70%. Despite an electrical efficiency of lower than 30%, the overall efficiency of the micro-CCHP system can reach as high as 90%. Consequently, such a compact system shows its feasibility in a domestic application.

Wang et al. [11] further carried out a case study using this micro-CCHP system. They operated this system in the CHP mode to meet the typical energy demands in the winter of Beijing, and in the CCHP mode to meet the ones in the summer of Hong Kong. In the CHP mode, the system either generates heat and electricity, or generates cold and electricity. In the CCHP mode, the electrical power output is fixed, and the ratio of heat to cold can be flexibly adjusted. Table 4.3 shows the multiply energy output and the overall efficiency in various cases. No matter how much the ratio of heat to cold is, the overall efficiency can maintain above 89%, revealing the flexibility and applicability of the DF-SOFC-based micro-CCHP system in a wide range of energy demands.

4.3.3 Costs of fuel cell-based micro-CHP systems

Even if fuel cell-based micro-CHP systems have been commercialized in many countries including Japan, Korea, etc., the price of these systems is still unsatisfactory. Followed by the target price of Department of Energy (DOE) in US, a number of estimates from academic institutions and industries believe that the prices of PEMFC-based stationary systems are possibly controlled at $\sim$\$1000 per kW by 2020. Limited by the cumulated installation, the actual prices of PEMFC-based and SOFC-based systems are \$24,000–57,000, 25–50 folds higher than the target price. In 2012, Staffell et al. [12] summarized the current price of the commercialized fuel cell-based stationary systems produced from

TABLE 4.3 Multienergy supply and system efficiency in different typical cases [11].

System configuration	Heat (W)	Cold (W)	Electricity (W)	System efficiency	
CHP mode	Heat + electricity	5124	0	922	89.2%
	Cold + electricity	0	5615	922	96.5%
CCHP mode	20% for Heat + 80% for Cold + electricity	1025	4329	922	93.4%
	50% for Heat + 50% for Cold + electricity	2740	2562	922	91.9%
	80% for Heat + 20% for Cold + electricity	4100	1096	922	90.3%

Japan, Europe, Korea, the United States, and tried to predict the gap between current price and the target. According to Staffell's summarization, the quantified prices for the fuel cell stationary systems with the cumulated installations were predicted according to the historical data fitting.

Fig. 4.13A shows that the fitting price curves over cumulated installations based on the historical data of the ENE-FARM systems (2004–11) and two other commercialized PEMFC systems [12]. Each dot denotes the price of a demonstrated system or commercialized system during one year and its current cumulated installation number. The fitting curves follow the equation below [12]:

$$P_n = P_{base} \times \left(\frac{Q_n}{Q_{base}} \right)^{-b} \tag{2}$$

where P_n and P_{base} are the price of the nth-generation unit and the reference unit, respectively. Q_n and Q_{base} are the cumulated installation number of the nth-generation unit and the reference unit, respectively. The exponential factor b is the experience parameter, that is, the slope of the fitting lines in Fig. 4.13A. The learning rate L is defined as $L = 1 - 2^b$. The price of the ENE-FARM systems in 2009 did not reduce, hence, the learning rate is smaller than the expected. The price in 2012 only slightly dropped, hence, the learning rate is only 15.0%. By fitting all the historical data in Fig. 4.13A, the expected price P_n can be expressed to be $P_n = \$33,624 \times (Q_n/1000)^{-0.251}$. The corresponding learning rate is equal to 16%. As this predication, the estimated cumulated installation number should be 1.2 billions to realize the target price of DOE, that is, $1000.

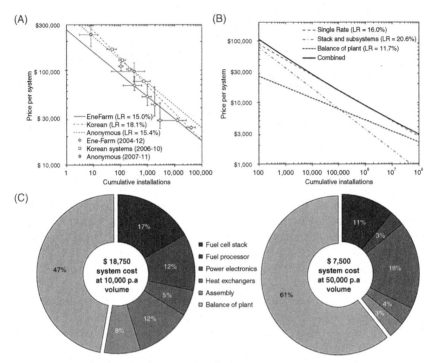

FIGURE 4.13 **(A) Price trend of current PEMFC-based micro-CHP systems with installations. (B) Price trend of different components with installations. (C) Target price distributions of ENE-Farm systems in 2004 and in 2011.** *(Modified from Ref. [12]. Copyright 2012 Elsevier Ltd.)*

To understand how the price of the fuel cell micro-CHP systems reduces, Fig. 4.13C shows two price distributions of the PEMFC-based micro-CHP systems estimated in 2004 and 2011 [12]. In 2004, the PEMFC system was sold at a price of $110,000. The price estimate in 2004 shows the system price will be $18,750 at an annual installation number of ~10,000. A latter price estimate was provided in 2011 by Strategic Analysis based on the assumption of a 50,000 annual installations. The price estimate considered the fuel cell stack, fuel processor, power electronics, heat exchangers, assembly, and BOP (involving pumps, valves, sensors, mass flow controllers, ejectors, additional structure, etc.), but the separate boiler and hot water tank were not counted. Both the price estimates reveal that the BOP plays the majority of the total costs, accounting for 47%–61%. The price for fuel cell stacks only accounted for 17% according to the earlier price estimate, while dropping to 11% in the latter price estimate. The significant differences between these two price estimate are the price of fuel processor, power electronics, and heat exchangers. The earlier estimate believed that the fuel processor and heat exchangers would account for a quarter of the total price, while the latter one showed that these two parts would only accounted for 7% and the power electronics would account for almost one fifth of the total price. As estimated, a PEMFC-based micro-CHP system involves at least 1000–2000 components and a solid oxide electrolysis cell(SOEC)-based one at least 200–600 components [12–14]. The numbers of the valves, pumps, blowers, and sensors in the key components of BOP are all more than 30, meanwhile, a number of pipe-work is also involved. More incentive was paid for the cost reduction in the fuel cell stack, while little for these little components that seem to be cheap and easy enough to obtain. Fig. 4.13B gives the fitting price lines of stack and subsystems, BOPs, and their combined systems. Here, a reference price of $18,750 fat a cumulated installation number of 50,000 was used according to the price estimate by ENE-FARM manufactory in 2014 as shown in Fig. 4.13C, and a learning rate of 16% was applied according to Fig. 4.13A. The learning rate of the stack+subsystems and BOPs are 20.6% and 11.7%, respectively according to the estimate in 2004 and 2011. By adding the price for the stack+subsystems and BOPs together, the total system price was drawn as the solid line in Fig. 4.13B. A high learning rate of the stack+subsystems leads to its predicted price falling with the cumulated installations, while the BOPs fall much slower. The combined line has a learning rate of 19.7% for the first system and 16.0% for the 100,000 installations. As the cumulated installations rise, the learning rate slows down due to the limitation of BOPs. This study offers a novel view on predicting the future price of the fuel cell micro-CHP systems. Based on the findings, Staffell et al. believed that a price goal of $3000–5000 by 2020 should be suitable and feasible for 1–2 kW-scale micro-CHP systems [12].

4.4 Hybrid fuel cell vehicle: Mobile distributed energy system

Hybrid fuel cell/battery vehicle system is a mobile fuel cell system, while fuel cell-based micro-CHP or CCHP systems are stationary DES. These two aspects seem to be two completely different application scenes. Like battery-powered electric vehicle can be energy storage carriers, fuel cell vehicles are also available power sources. Flexible plug-and-play power sources are beneficial for decentralizing and reducing the transmission losses. The "Green Village" programme from Delft University of Technology proposed a novel distributed micro-CCHP tri-generation system powered by the fuel cell vehicles as Fig. 4.14 shows [15]. This power plant locates in a parking lot and consists of three regions: Hydrogen production region, parking region and pump station region. In the hydrogen production region, natural gas and water are fed and reacting with each other to produce hydrogen for the pump station region. SOFC stacks are alternatively installed in the hydrogen production region, hence, the hydrogen production region can operate as a micro-CCHP system for tri-generation of electricity, heat and cold. The pump station region is served as a small-scale hydrogen refueling station, mainly aims to refuel hydrogen for the passing fuel

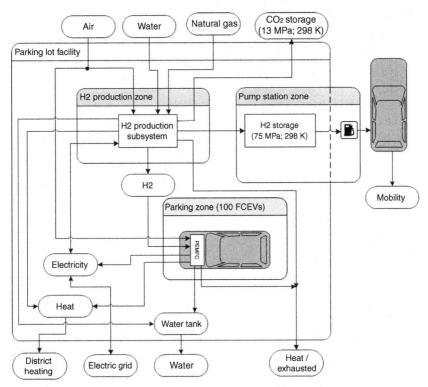

FIGURE 4.14 Schemetic of the "Car Park Power Plant" tri-generation system in "Green Village" programme. *(Modified from Ref. [15]. Copyright 2016 Elsevier Ltd.)*

cell vehicles. In the parking region, the owners can park their fuel cell vehicles here and leave. During this time, the parked fuel cell vehicles plug in the hydrogen production region, and act as the fuel cell module in the common fuel cell-based micro-CCHP systems to generate electricity, heat and hot water.

Fernandes et al. [15] carried out a comprehensive exergy analysis for such a "Car Park Power Plant" system. To obtain a high efficiency and lower CO_2 emission, they compared the exergy efficiency before or after replacing the catalytic reformer with SOFC, meanwhile, the effect of the installation of CO_2 capture and storage device (CCS) on the exergy efficiency was also discussed. Simulation results indicate that replacing the catalytic reformer with SOFC can remarkably enhance the exergy efficiency by at more than 20%. When CCS is integrated, the enhancement in the exergy efficiency rises to 30%. When the catalytic reformer is used, the integration of CCS leads to a dramatic drop of more than 7% in the exergy efficiency. After the catalytic reformer is upgraded to SOFC, this drop is significantly inhibited, reducing to only 2%. Overall, the SOFC-integrated "Car Park Power Plant" system can realize an overall tri-generation efficiency of 60% based on the exergy analysis [15].

4.5 Summary

In this chapter, we offered some case studies on hybrid fuel cell systems applied for fuel cell vehicles and micro-CHPs or CCHPs. Both PEMFC-based and SOFC-based fuel cell systems were demonstrated when applied in the automotive field and DESs. The key indicators of these fuel cell systems were compared and discussed. Despite it seems apparent that PEMFC is suitable for the automotive application and SOEC is suitable for micro-CHPs or CCHPs, a complementary integration and appropriate design of the hybrid system is possible to offset some apparent inapplicability in a certain application scene. Finally, a novel concept named "Car Park Power Plant" is briefly introduced, which is potential to reduce the transmission losses and promote the decentralization of the future energy systems.

References

[1] Department of Energy. Comparison of Fuel Cell Technologies. Accessed to Apr. 3, 2019. <https://www.energy.gov/eere/fuelcells/comparison-fuel-cell-technologies>.

[2] S. Giddey, S.P.S. Badwal, A. Kulkarni, C. Munnings, A comprehensive review of direct carbon fuel cell technology, Prog Energy Combust 38 (2012) 360–399.

[3] K.S. Jeong, O.B. Soo, Fuel economy and life-cycle cost analysis of a fuel cell hybrid vehicle, J Power Source (2002) 58–65.

[4] Nissan announces development of the world's first SOFC-powered vehicle system that runs on bio-ethanol electric power. 2016, accessed to Apr. 11, 2019. <https://nissannews.com/es/nissan/ lac/releases/nissan-announces-development-of-the-world-s-first-sofc-powered-vehicle-system-that-runs-on-bio-ethanol-electric-power?mode=print>.

[5] J.M. Ogden, M.M. Steinbugler, T.G. Kreutz, A comparison of hydrogen, methanol and gasoline as fuels for fuel cell vehicles: implications for vehicle design and infrastructure development, J Power Source (1999) 143–168.

[6] Y. Bessekon, P. Zielke, A.C. Wulff, A. Hagen, Simulation of a SOFC/Battery powered vehicle, Int J Hydrogen Energy 44 (2019) 1905–1918.

[7] V. Alfonsin, R. Maceiras, A. Cancela, A. Sanchez, Modelization and simulation of an electric and fuel cell hybrid vehicle under real conditions, Eur J Sustain Dev (2015) 4.

[8] R. Napoli, M. Gandiglio, A. Lanzini, M. Santarelli, Techno-economic analysis of PEMFC and SOFC micro-CHP fuel cell systems for the residential sector, Energy Build 103 (2015) 131–146.

[9] A. Adam, E.S. Fraga, D.J.L. Brett, Options for residential building services design using fuel cell based micro-CHP and the potential for heat integration, Appl Energy 138 (2015) 685–694.

[10] M. Horiuchi, S. Suganuma, M. Watanabe, Electrochemical power generation directly from combustion flame of gases, liquids, and solids, J Electrochem Soc 151 (2004) A1402.

[11] Y. Wang, Y. Shi, M. Ni, N. Cai, A micro tri-generation system based on direct flame fuel cells for residential applications, Int J Hydrogen Energy 39 (2014) 5996–6005.

[12] I. Staffell, R. Green, The cost of domestic fuel cell micro-CHP systems, Int J Hydrogen Energy 38 (2013) 1088–1102.

[13] R. Tanaka, Update on residential fuel cells demonstration and related activities in Japan, Fuel cells KTN: fuel cells for distributed generation.

[14] T. Ueda, Japan's approach to the commercialization of fuel cell/hydrogen technology, 7th steering committee meeting, International Partnership for the Hydrogen Economy, Sao Paulo, Brazil, 2007.

[15] A. Fernandes, T. Woudstra, A. van Wijk, L. Verhoef, P.V. Aravind, Fuel cell electric vehicle as a power plant and SOFC as a natural gas reformer: an exergy analysis of different system designs, Appl Energy 173 (2016) 13–28.

Chapter 5

Stabilization of intermittent renewable energy using power-to-X

Yu Luo[a], Yixiang Shi[b] and Ningsheng Cai[b]

aNational Engineering Research Center of Chemical Fertilizer Catalyst (NERC-CFC), Fuzhou University, Fujian, China; bDepartment of Energy and Power Engineering, Tsinghua University, Beijing, China

5.1 Introduction

As evaluated, DHS needs approximately 15%–20% of the annual loads to meet 2–3 month of storage [1,2]. Other than pumped-hydro and compressed air energy storage, storing energy in fuels is the only energy storage technology applicable for seasonal energy storage [1,3]. Power-to-X technology can convert surplus renewable power into energy-dense fuels through electrolyzers and sequent fuel synthesis processes so as to store renewable energy sources with zero self-discharge losses. Power-to-gas (PtG) and power-to-liquid (PtL) are two main power-to-X routes. The most common electrolyzers are water electrolyzers, where H_2O is electrochemically reduced to H_2. High-purity H_2 produced by electrolyzers can directly supply for fuel cells and hydrogen power plants, or be transported to hydrogen refueling station. Essentially, PtG or PtL is to store renewable energy in hydrogen-rich fuels or chemicals. Even though power-to-hydrogen is the most common PtG pathway, hydrogen storage is still a challenge. High diffusibility, buoyancy, and activity of H_2 make it difficult to store [4]. Generally, H_2 needs to be pressurized to 350–700 bar for storage and transportation, which causes an increase of risk level. Some hydrogen storage technologies such as cryo-compression, metal-hydrides, and adsorption are in development, but still not cost-efficient. Storing hydrogen in chemical hydrides reveals lower costs and higher technical maturity than other hydrogen storage technologies.

Hydrocarbons such as methane, methanol, gasoline are usually selected as the storage media, which can also recycle CO_2 emitted from fossil fuel-based power plants or chemical industries. CO_2 can react with H_2 generated from H_2O electrolyzers to produce various hydrocarbons, or CO_2 and H_2O are co-electrolyzed through solid oxide electrolysis cells (SOEC) to syngas

Hybrid Systems and Multi-energy Networks for the Future Energy Internet.
http://dx.doi.org/10.1016/B978-0-12-819184-2.00005-5

113

(H_2 + CO mixture) for sequent hydrocarbon production, even directly to produce target hydrocarbons. Hydrocarbons are available for their current applications such as power generation, refueling station, multiple chemicals, and jet fuels. Thus, current devices and infrastructures can avoid specific reformation (re-designed for utilizing H_2). Particularly, liquid hydrocarbons are more convenient for transportation, hence, can be easier to be sold to downstream markets.

In this chapter, we will give some case studies of typical PtG or PtL systems, including power-to-H_2 (PtH), power-to-syngas (PtS), power-to-methane (PtM), power-to-methanol (PtMe), power-to-F-T fuels, etc. The technical characteristics, energy efficiency, ecnomic feasibility, and their future markets are all comprehensively discussed.

5.2 Power-to-gas systems

5.2.1 Power-to-H_2 for hydrogen production

As introduced in Chapter 3, PtH uses electrolyzers to produce H_2 from H_2O. The common electrolyzers include alkaline electrolysis cells (AEC), proton exchange membrane electrolysis cells (PEMEC), and SOEC. Fig. 5.1A shows the working principles of these electrolyzers. These electrolyzers transfer different ions, leading to the difference in the electrolyte materials. AEC uses alkaline solutions such as potassium hydroxide (KOH) or sodium hydroxide (NaOH) solution as the electrolyte to transfer hydroxide ions. PEMFC uses perfluorosulfonic acid (PFSA) polymer, for example, Nafion®, as the electrolyte membrane to transfer proton. SOFC uses solid oxide ceramic materials, for example, yttrium stabilized zirconia (YSZ), as the electrolyte to transfer oxygen ions. These electrolyte materials determine the preferred operating temperature of these electrolyzers. AEC and PEMEC prefer to operate below 100°C, while SOEC generally operates at 500–1000°C. Higher operating temperature can reduce the theoretical electricity consumption and promote reaction kinetics. Fig. 5.1B compares the polarization curves of AEC, PEMEC, and SOEC. A polarization curve cantains the information related to both thermodynamics and reaction kinetics. The applied voltage at zero current, that is, open-circuit voltage (OCV), reveals the minimum electricity consumption of an electrolyzer. As current density increases, the applied voltage rises. At a fixed current density, the hydrogen production rate is fixed when no secondary reactions. The difference between the applied voltage and OCV at a fixed current density, that is, polarization voltage or overpotential, is up to the reaction kinetics. Overall, higher the applied voltage at a fixed current density, lower is the electricity consumption of an electrolyzer. Fig. 5.1B reveals that the OCV of AEC and PEMEC are larger than 1.4 V, while that of SOEC is only 0.95 V. This means that the minimum electricity consumption of AEC or PEMFC is at least 47% higher than that of SOEC. The slope of the polarization curves can reflect the polarization resistance. SOEC and PEMEC have similar

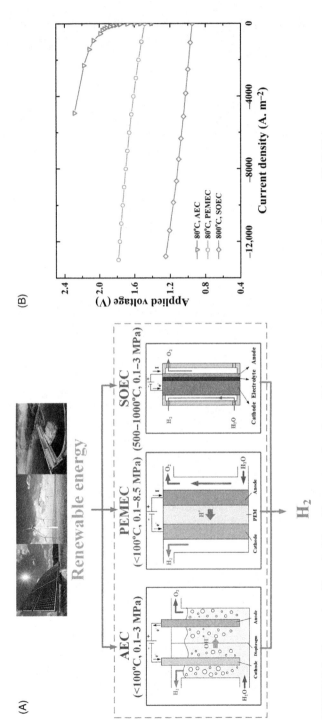

FIGURE 5.1   Types and working principles of the common electrolyzers (A), and their typical electrochemical performance (B) [5–7].

polarization resistance, that is, similar polarization voltage at a fixed current density, while AEC has a much higher polarization resistance than the other two electrolyzers. Consequently, to realize a current density of -4000 A m^{-2} (corresponding to a hydrogen production rate of 1.67 Nm^3 h^{-1} m^{-2}), AEC requires an applied voltage of ~2.2 V (consuming ~8.4 kW m^{-2} effective reaction area), PEMEC requires an applied voltage of ~1.6 V (~6.4 kW m^{-2}), and SOEC requires an applied voltage of ~1.03 V (~4.1 kW m^{-2}). The electricity consumption of AEC is 30% higher than that of PEMEC, and over twice of that of SOEC. SOEC can save at least 36% electricity than low temperature electrolyzers. However, this estimate doesn't consider the other energy consumption from the auxiliary equipment. Particularly, a PtH system using SOEC requires more heating devices, thus, a comprehensive energy analysis in system level is in need.

5.2.1.1 Efficiency and energy consumption

In our group's previous study [8], we carried out a comprehensive exergy analysis of various PtM systems. Despite we didn't analyze the energy efficiency of a separate PtH system, we compared the overall energy efficiency and energy distribution in hybrid PtH + methanation systems using different electrolyzers (including AEC, PEMEC, and SOEC). Fig. 5.2A reveals the system layout of a hybrid PtH + methanation system discussed in Ref. [8]. The upper part in Fig. 5.2A is the PtH subsystem using SOEC, and the lower part is the sequent processes including methanation, CO_2 removal, etc. In the systems, heat exchanges were used to recover heat from the outlets of SOEC stack, and heaters for preheating inlet gas to the operating temperature of SOEC. The produced hydrogen was compressed by a compressor for the sequent methanation reactor. Here, we set the heat exchange effectiveness to 90%, and the efficiency of compressors and turbines to 80%. When the PtH subsystem used AEC or PEMFC, the heating devices can be removed. Liquid water can be directly fed into the electrolyzers, and the produced H_2 from the electrolyzers was fed into the methanation reactor after compression.

Fig. 5.2B gives the exergy efficiency of the hybrid PtH + methanation systems using AEC, PEMEC, and SOEC with current density (proportional to hydrogen production rate). Here, the performances of these three electrolyzers are consistent with the ones in Fig. 5.1B. AEC and PEMEC operate at 80°C, and SOEC at 800°C. As electrical power is applied to electrolyzers, the exergy efficiency of PEMEC rapidly increases with current density to more than 40% (current density >500 A m^{-2}), while that of SOEC and AEC need at least 1000 A m^{-2} even higher to obtain an exergy efficiency of 40%. This reveals that PEMEC can maintain high efficiency in a wide load range, hence, has higher capability of load adjustment. As for AEC-based and SOEC-based systems, different reasons lead to their low efficiency at low current density. For SOEC-based system, it is less efficient than PEMEC-based system at a current density of lower than 3400 A m^{-2} due to high operating temperature. At low current

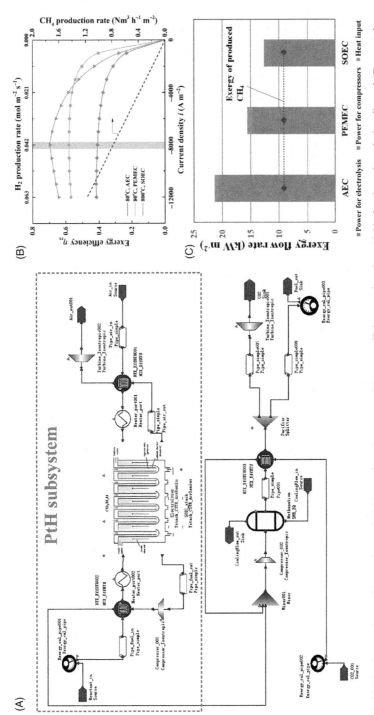

FIGURE 5.2 (A) System layout of a hybrid PtH + methanation system; (B) exergy efficiency and methane yield of various electrolysis cells; and (C) comparison in energy distribution. *(Modified from Ref. [8]. Copyright 2018 Elsevier Ltd.)*

density, the applied voltage of SOEC is below thermal neutral voltage, that is, TNV. Consequently, SOEC is in endothermic mode and requires extra heat input to support enough high operating temperature. As current density rises, SOEC switches from endothermic mode to exothermic mode and becomes temperature sustainable. SOEC unit has higher electrical efficiency than PEMEC one. As a result, SOEC-based system exceeds PEMEC-based one and reaches the peak efficiency (70%) at 8000 A m^{-2}, which is 11% higher than the peak efficiency of the PEMEC-based system. As for the AEC-based system, its low efficiency is attributed to the high polarization resistance of AEC unit. The AEC-based system has the lowest exergy efficiency among these three systems, only less than 42%. When current density is ~8000 A m^{-2}, these three systems almost reach the highest efficiency. Fig. 5.2C gives more details of energy consumption distributions in these systems at 8000 A m^{-2}. In these systems, we set the CH$_4$ production rate to be the same. The exergy flow shows that AEC-based system consumes 22 kW energy per m^2 effective reaction area, the most energy among these three systems. PEMEC-based system consumes 16 kW m^{-2}, 27% lower than AEC-based system. SOEC-based system has the lowest energy consumption, only 12.5 kW m^{-2}. This result accords with the exergy efficiency in Fig. 5.2A. In the AEC-based system, the electricity consumed by AEC reaches over 20 kW m^{-2}, accounting for over 94% of the overall energy consumption. Similarly, in the PEMEC-based system, the electricity consumed by PEMEC drops to around 14 kW m^{-2}, accounting for 93% of the overall energy consumption. When SOEC is integrated into the system, the electricity consumption is reduced to 11 kW m^{-2}, and the corresponding ratio in the overall energy consumption decreases to 86%. In other words, at least 86% of energy consumption is inevitably used for powering the electrolyzers no matter which type of electrolyzer is selected. At the optimal current density, SOEC operates above TNV, hence, requires no extra heat input. AEC and PEMEC have higher heat requirement than SOEC because of the heat demand in the latter methanation reaction. Low temperature of AEC and PEMEC results that these two systems need extra energy input to upgrade the heat generated from AEC and PEMEC.

5.2.1.2 Life-cycle green-house gas emission

Based on AEC and PEMEC, Zhang et al. [9] compared the life-cycle green-house gas (GHG) emission of the PtH subsystems powered by various electricity sources with conventional fossil fuel-based hydrogen production technologies, as Fig. 5.3 shows. In Zhang's study, they considered the electricity sources including wind power, photovoltaic power (PV), average power supply from the Swiss grid (mixed electricity sources, the majority from hydropower and nuclear power), and average power supply from the European grid (the majority from fossil fuels). Considering the intermittence and fluctuation of renewable energy sources in Switzerland, Zhang et al. added error bars into the estimated GHG emission. The limited penetration in the Swiss grid and the European grid leads

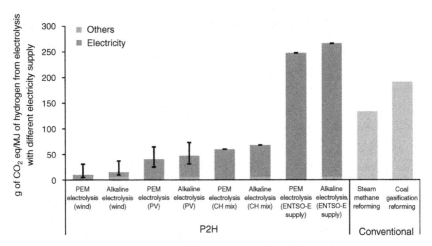

FIGURE 5.3   Life-cycle CO_2 emissions of PtH using different electrolyzers or driven by different electricity sources, and those of conventional fossil fuel-based H_2 production. *(From Ref. [9]. Copyright 2016 Elsevier Ltd.)*

in the ignorable error bars in these two cases. Even though wind power and PV are both renewable energy sources, PV shows higher average GHG emission as well as larger fluctuation than wind power. When it comes to the PEMEC-based PtH system, the PtH system powered by wind emits about 10 CO_2 equivalent green-house gas per MJ (with a fluctuation range from 4 to 31 g CO_2 equiv. per MJ), only a quarter of the GHG emission from PV power (41 g CO_2 equiv. per MJ, fluctuating from 25 to 66 g CO_2 equiv. per MJ). The fluctuation range of the wind-driven PtH system is 27 g CO_2 equiv. per MJ, 31% smaller than the PV-driven PtH system. As for the mix electricity sources from the Swiss grid, the large penetration of hydropower and nuclear power results in a GHG emission of 55 g CO_2 equiv. per MJ, higher than the wind power-driven PtH system and equivalent to the PV power-driven PtH system. However, when large-scale electricity from the fossil fuel-based European grid, the GHG emission dramatically soars up by at least 3.5 folds to ~250 g CO_2 equiv. per MJ. This value is even double of the conventional hydrogen production from steam methane reforming (SMR) and 30% higher than the one from coal gasification reforming (CGR). This is because fossil fuel-to-power plus PtH causes extra efficiency losses rather than direct fossil fuel-to-H_2 (conventional pathway). Therefore, PtG or PtL is not environmentally friendly at all when fossil fuel-based electricity is applied. In a whole life cycle, the AEC-based PtH system emits slightly more GHG than the PEMEC-based PtH system. When the electricity sources come from renewable energy sources or clean energy sources like wind, solar PV, hydropower, or nuclear, higher GHG emission of the AEC-based PtH system is attributed to the device manufacture (*Green columns*, ~5 g CO_2 equiv. per MJ). When the electricity sources come from the fossil fuel-dominated grid, extra GHG (~6%)

is emitted by the AEC-based PtH system because of the relatively low efficiency of AEC. Consequently, in the PtH system, the contribution of electricity input in the total GHG emission is dominant. Nevertheless, this contribution becomes comparable to the one caused by the electrolyzer manufacture when the PtH system is driven by wind power or other clean energy sources with a GHG emission as low as ~ 10 g CO_2 equiv. per MJ.

5.2.2 Power-to-syngas via H_2O/CO_2 co-electrolysis

Besides PtH and PtM, PtS is also an optional PtG pathway. Syngas is an important chemical feedstock for the production of methane, methanol, gasoline, diesel, and other hydrocarbons. High temperature H_2O/CO_2 co-electrolysis using SOEC can directly convert H_2O and CO_2 into syngas with a regulable ratio of H_2 to CO, further offering the desired syngas for the sequent industrial productions of methane, methanol, and F-T fuels. H_2O/CO_2 co-electrolysis is currently the most feasible and promising direct-PtS technology even though the technical maturity is still not enough. However, this technology offers us a novel pathway to recycle or fix the emitted carbon.

In our group's previous study [8], we compared the overall exergy efficiency of various PtM systems, including the system integrating PtS and methanation, and the previously mentioned PtH + methanation system. When we compared these two systems, SOEC was selected as the electrolyzer for the production of both H_2 and syngas. Although we didn't analyze the energy efficiency of a separate PtH system and a separate PtS system, the overall exergy efficiency of these hybrid systems can reveal some distinguish features of the PtS technology using solid oxide H_2O/CO_2 co-electrolysis cells. On the one hand, the C=O bonds are hard to break, leading to a worse performance of H_2O/CO_2 co-electrolysis than H_2O electrolysis. On the other hand, CO_2 is mixed with H_2O into the SOEC, enriching the reactants for the electrolysis reaction. Fig. 5.4A reveals the exergy efficiency of the hybrid PtH + methanation system and the hybrid PtS + methanation system at an H/C ratio of 10.54. Ignoring the Faraday efficiency losses, the produced CH_4 from these two systems should be close because of the quasiequilibrium in the methanation reactors. These two systems both reaches the maximum exergy efficiency in a current density range from -5500 to -6000 A m^{-2}, in which the overall heat exergy flows balance (approximately equal to 0). The peak value of the exergy efficiency in the hybrid PtH + methanation system is equal to 70.1% (obtained at 5500 A m^{-2}), while that in the hybrid PtS + methanation system is 69.5% (obtained at 6000 A m^{-2}). This result reveals that the PtS-integrated system has a slightly lower exergy efficiency than the PtH-integrated system, but performs better at relatively high current density (>7000 A m^{-2}). Higher current density for the peak exergy efficiency in the PtS-based system is attributed to more abundant reactants (adding CO_2 in the inlet gas) and higher TNV. H_2O electrolysis has a TNV of 1.28 V, while CO_2 electrolysis has one of 1.47 V. Thus, the TNV of H_2O/CO_2 co-electrolysis

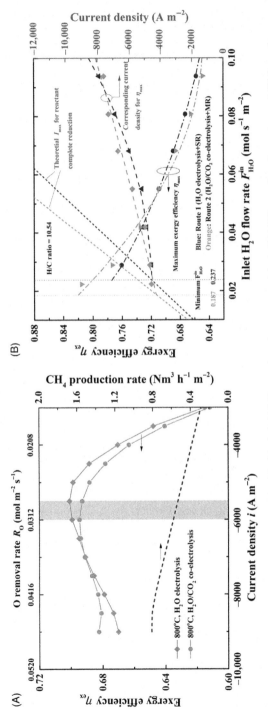

FIGURE 5.4   (A) Exergy efficiency and methane yield of the hybrid PtH + methanation system and the hybrid PtS + methanation system and (B) the peak exergy efficiency and the corresponding current density with inlet flow rate. (*Modified from Ref. [8]. Copyright 2018 Elsevier Ltd.*)

should be between 1.28 and 1.47 V. Fig. 5.4B reveals the peak exergy efficiency and the corresponding current density with inlet steam flow rate at an H/C ratio of 10.54. The decrease in the inlet reactant flow rate reduces the heat for heating inlet gas and increases the reactant conversion due to a fixed current density. This leads to an increase in the peak exergy efficiency and a drop in the current density for the peak exergy efficiency. By fitting these calculated dots, we tried to predict the potential limitation of the exergy efficiency. Assuming a Faraday efficiency of 100%, the inlet reactant flow rate is proportional to the theoretical current density limitation. The intersection between the fitting current density for the peak efficiency and the limiting current density reveals that the potential maximum for the peak exergy efficiency could be 77.3% for H_2O electrolysis (feeding 0.19 mol s^{-1} m^{-2} steam), while 82.0% for H_2O/CO_2 co-electrolysis (feeding 0.24 mol s^{-1} m^{-2} steam). Consequently, the hybrid PtS-based system using H_2O/CO_2 co-electrolysis has a higher peak efficiency at a reduced inlet reactant flow rate, revealing its applicability at high fuel production rate.

Fu et al. [10] from European Institute for Energy Research (EIFER) carried out a comprehensive economic study on a PtS system. In this system, Fu et al. considered the SOEC operating at 1.1–1.343 V with a maintenance of 4% and an ASR of 0.25–1.25 Ω cm^2. The sensitivity of the SOEC investment cost (varying from 500 to 5000 € m^{-2} effective cell area) and lifetime (varying from 5000 to 8000 h) were also evaluated. In FU's estimated system [10], the capacity of the SOEC stack is around 1.5 MW, the feeding steam flow rate is 6.0 t (~331 kmol) for 1 day and the CO_2 flow rate is 6.9 t (~157 kmol) per day. As estimated, the produced syngas in the base case is 5.1 t (0.44 t CO + 0.67 t H_2) with an H/C ratio of 4.23, meanwhile, 7.8 t pure oxygen is a byproduct. The overall energy inputs are 1.61 MW, consisting of 1544 kW electricity, 22 kW low temperature heat, and 46 kW high temperature heat. The low temperature heat comes from the nuclear reactor, while the high temperature heat from the electricity. Consequently, 98.6% of the total energy consumption is the electricity consumption in Fu's PtS system. The estimated cost for syngas production in the base case is 0.775 € kg^{-1}, where the cost of the consumed electricity accounts for 57%, the feedstock cost for 32% (high CO_2 price, ~160 €), and the capital maintenance only for 9%. The energy input and feedstock input account for 89% of the total costs. Fig. 5.5A gives the sensitivity of the syngas cost to the SOEC performance and lifetime [10]. Generally speaking, an increase in SOEC performance can reduce the required reaction area and further lower the stack cost. For the SOEC lifetime, there is a critical lifetime existing for the SOEC stack. When the lifetime is higher than the critical value, the syngas cost only reduces slightly with the lifetime increasing. When the lifetime is lower than the critical value, the syngas cost significantly soars with the lifetime shortening. When ASR = 0.25 Ω cm^2, the critical lifetime is around 10,000 h. When the SOEC performance is dropped to ASR = 1.25 Ω cm^2, the critical lifetime is doubled. As the SOEC ASR reduces, the syngas cost linearly drops. Higher the lifetime is, slower the drop of the syngas cost with ASR. When the lifetime

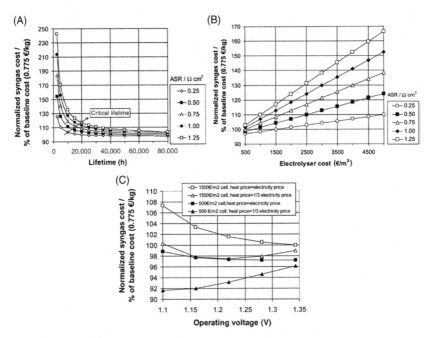

FIGURE 5.5 (A) Effect of the SOEC ASR and lifetime on the cost for syngas production; (B) effect of the SOEC cost on the cost for syngas production; and (C) effect of the operating voltage on the cost for syngas production. *(From Ref. [10]. Copyright 2010 Royal Society of Chemistry.)*

can be ensured over 20,000 h, the syngas cost with an ASR of 1.25 Ω cm^2 is only 17% higher than the one with an ASR of 0.25 Ω cm^2. As a result, we don't need to go to the one extreme of the SOEC performance or lifetime. An target lifetime of 20,000 h and a target ASR of 0.75 Ω cm^2 can offer acceptable syngas cost (at most 9% higher than the one with a lifetime of 80,000 h and an ASR of 0.25 Ω cm^2). Fig. 5.5B reveals the sensitivity of the syngas cost to the investment cost of SOEC at various ASR. The syngas cost is proportional to the investment cost of SOEC, while the sensitivity can reduce with the enhanced SOEC performance (the ASR reducing). When ASR = 0.25 Ω cm^2, an increase in the investment cost of SOEC by one order of magnitude (from 500 to 5000 € m^{-2}) only results in the syngas cost rising by less than 15%. However, when ASR rises to 1.25 Ω cm^2, the same increase in the investment cost of SOEC leads to the syngas cost rising by 64%. Reducing the cost of SOEC stacks with an ASR of 1.25 Ω cm^2 to 1000 € m^{-2} leads to the same syngas cost as reducing the ASR of SOEC stacks with 5000 € m^{-2} to 0.25 Ω cm^2. The product of the SOEC ASR and SOEC cost determines their effects on the overall syngas cost. Consequently, a trade-off should be made between the SOEC performance and the SOEC cost. Fig. 5.5C shows that the syngas cost with the operating voltage when considering two different SOEC prices (500 and 1500 € m^{-2})

and two different high temperature heat prices. As the operating voltage is lower than TNV, SOEC demands more heat to support the operating temperature and less electricity for electrolysis. At 1.1 V, 18% of electricity can be saved by high temperature heat supply. When considering the heat price is equal to electricity price, the syngas cost at low operating voltage is not cost-effective, even more expensive than above TNV. Too high operating voltage requires too much electricity consumption, while too low operating voltage increases the required reaction area of SOEC. When the heat price drops to one-third of the electricity price, the syngas cost becomes relatively high at a high operating voltage. At low operating voltage, the syngas cost is also relatively high at a high SOEC cost ($1500 €$ m^{-2}), while the PtS system becomes cost-effective at a low SOEC cost ($500 €$ m^{-2}). When the SOEC cost is relatively high, the SOEC was recommended to operate at TNV for better thermal management and lower syngas cost. Furthermore, Fu et al. [10] also considered the effects of various energy source inputs on the syngas price. When wind power replaces nuclear power for the PtS system, the investment cost increases due to the intermittence of wind energy (larger installed capacity). As Fu's summarized representative price, the nuclear power cost was ~$56 €$ MWh^{-1}, the onshore wind power cost was ~$90 €$ MWh^{-1}, and the offshore wind power cost was ~$150 €$ MWh^{-1}. Correspondingly, the syngas cost is expected to be $1.27 €$ kg^{-1} for the use of onshore wind power and $1.65 €$ kg^{-1} for using offshore wind power. When the cost of the wind power can be dropped to $45 €$ MWh^{-1}, the syngas cost can be comparable with the one using nuclear power [10].

5.2.3 Power-to-methane for integrating with the natural gas networks

Conventional PtM systems are similar to the one in Fig. 5.2A, integrating electrolyzers with the sequent methanation. Compared with the PtH system, the PtM system has an extra efficiency loss due to strongly endothermic methanation process. However, hydrogen storage is still costly and of relatively low safety. In contrast, methane, as the main content of natural gas, can be directly sent into the natural gas network for easier and safer transportation. Besides, methane as the fuel has been widely used for current energy devices like internal combustion engines (ICE), gas turbines, burners, etc., hence, the use of methane requires no modification design on these devices. Currently, there are three pathways available for the PtM conversion.

5.2.3.1 PtM vs PtH

Typically, the methanation reactor operates at 200–500°C and 20–25 bar [11]. When using low temperature electrolyzers, the PtM system needs to integrate a PtH subsystem with the methanation reactor. The methanation reactor releases a large amount of heat with higher temperature, which is hard to be utilized in the PtM system due to the absence of high temperature heat demand. If the heat is unable to be used for the nearby plants, the wasted heat results in the efficiency loss of the PtM system as high as 16% compared with the PtH system [12]. As reported, the first industry-scale PtM plants built by ETOGAS for Auti AG in Werlte, Germany has a power load of ~6 MW_e, but the overall efficiency is only ~54% [13]. When using high temperature SOEC, the PtM system can utilize the heat generated from the methanation to realize a more efficient heat integration and reduce the efficiency losses, even the SOEC and methanation reactor are potential to be integrated into one reactor. A better heat integration can significantly reduce the TNV of the SOEC reactor, hence, reduce high-grade heat waste. Thus, an SOEC-based system can significantly decrease the efficiency loss when turning PtH to PtM.

Zhang et al. [9] compared the estimated environmental impacts of the PtH and PtM technologies, as Fig. 5.6 shows. For a fair comparison, the electrolyzer used in the PtH or PtM system is uniformed to the PEMEC with a capacity of 100 kW_e. PV power and wind power are selected as the electrical energy sources, respectively. The GHG emission reduction for the PtH system is evaluated by comparing the PtH system with conventional H_2 production from fossil fuels (SMR or CGR), while the one for the PtM system is based on the comparison with natural gas for the application of the vehicle fuels. When the PtH systems are driven by PV power instead of the conventional fossil fuels, the GHG emission can be reduced by at least 200 g CO_2 equiv. per kWh. If the compared fossil fuel is coal, this reduction can be higher, that is, 330 g CO_2 equiv. per kWh. Wind power reveals a 70 g CO_2 equiv. per kWh higher GHG emission reduction than PV power. The GHG emission reduction can increase to 270–400 g CO_2 equiv. per kWh. The lower boundary is corresponding to the comparison with SMR, and the upper one is corresponding to the comparison with CGR. The PtH

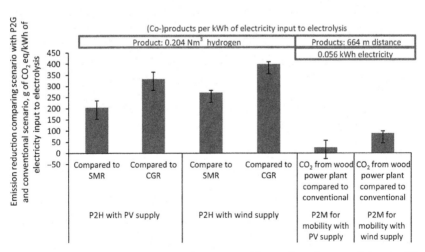

FIGURE 5.6   Life-cycle CO_2 emissions of PtH using different electrolyzers or driven by different electricity sources, and those of conventional fossil fuel-based H_2 production. *(From Ref. [9]. Copyright 2016 Elsevier Ltd.)*

system uses a PEMEC stack to generate 0.204 Nm^3 hydrogen per kWh. When it comes to the PtM system, the application scene changes. The PtM system was used to supply vehicle fuels for replacing natural gas. For 1 kWh electricity consumption, the methane produced from the PtM system can support a vehicle to run a distance as far as 0.66 km and generate 0.056 kWh electricity in the meantime. The GHG emission reduction of the PtM system is less than that of the PtH system, only 25 g CO_2 equiv. per kWh for the PV power-driven case and 88 g CO_2 equiv. per kWh for the wind power-driven case. This is resulted from the difference in the compared reference application. The PtH system can avoid the use of the fossil fuels in the hydrogen production processes, thus, significantly reduces the CO_2 emission during SMR or CGR. As for the PtM system, the GHG emission reduction refers to the reduced GHG emission in a wood power plant caused by integrating CO_2 capture and PtM into the conventional system. The driving distance is uniformed in this case. In this system, CO_2 is a basic feedstock, and comes from a fossil fuel-based power plant. Despite CO_2 is recycled in this system, methane emits CO_2 in the end user, that is, vehicles. This part of CO_2 emission was not considered to be recycled. Thus, this system inevitably emits an amount of CO_2. Consequently, the GHG emission reduction of the PtM system is less than that of the PtH system. If we only consider the net GHG emission reduction only in the PtM system, the recycled CO_2 in the PtM system offsets the produced CO_2 from the end-use of the methane to reveal a higher GHG emission reduction than the current comparison case.

5.2.3.2 Operating condition optimization

A suitable operating condition is significant for maximizing the CH_4 production and minimizing the energy consumption. Our previous study [8] studied the effect of the applied current density and inlet H/C ratio on the overall efficiency in a system integrating water electrolyzers and methanation reactors. The effect of current density has been discussed in Section 2.1.1 and Fig. 5.2. The peak efficiency exists as the current density varies, particularly for the SOEC-based PtM system. The peak efficiency is obtained when the total heat effect turns from endothermic to exothermic, that is, in the system-level thermal neutrality. As the H/C ratio rises and inlet water conversion fixes at 80%, the methane production keeps constant at H/C ratio <10 and decreases at a H/C ratio of higher than 10. On the contrary, the outlet molar fraction of H_2 keeps lower than 10% at H/C ratio <10, while remarkably increases with H/C ratio when H/C ratio is higher than 11. Due to the rise in H_2 content, the overall efficiency slightly increases at low H/C ratio (<10) and soars up with the H/C ratio increasing to higher than 10. Considering the allowed H_2 content limitation of 15% in volume, we optimized the H/C ratio to 10.5. At the optimal H/C ratio, the PtM system integrating water electrolyzers and methanation reactor can offer an overall efficiency of 71% (exergy efficiency) and a methane production rate of 0.81 $Nm^3 \, h^{-1} \, m^{-2}$.

Stempien et al. [14] carried out a study on the operating condition optimization of an PtM system integrating a solid oxide H_2O/CO_2 co-electrolysis cell stack and methanation reactor. Effects of the steam-to-carbon dioxide ratio (SCR), temperature and pressure on the energy efficiency and the methane production were evaluated, as Fig. 5.7 shows. In Stempien's reference case, the SOEC operated at 800°C, the methanation reactor at 400°C and the whole system operates at 3 bar. These three lines denote that the SOEC operates at 0.5, 1, and 1.5 A cm^{-2}. Fig. 5.7A shows the energy-to-methane efficiency *(dash line)* and electricity-to-methane efficiency *(solid line)* at various SCR. As the SCR varies, there is a peak efficiency existing for the PtM system, moreover, the energy-to-methane efficiency and the electricity-to-methane efficiency reach the peak almost at the same SCR. The highest peak electrical-to-efficiency efficiency can be obtained at SCR = 7 and current density = 0.5 A cm^{-2}, equal to 81%. However, the corresponding energy-to-methane efficiency is the lowest, only 43%. As the applied current density rises, the peak electricity-to-efficiency efficiency decreases while the peak energy-to-methane efficiency increases. At 1.5 A cm^{-2}, the peak energy-to-methane efficiency reaches 60%, but the corresponding electricity-to-efficiency efficiency is only 72%. Higher current density increases the heat generated from the SOEC, hence, reducing the electricity-to-methane efficiency but reducing the external heat demand. Consequently, both efficiencies become closer to each other. Fig. 5.7B shows the methane production rate and CO_2-to-CH_4 conversion at various SCR. The corresponding SCR for the peak efficiency is exactly the optimal SCR for the peak methane production rate. As the applied current density rises, the peak methane production rate increases from 0.6 to 1.5 Nm^3 h^{-1} m^{-2}. At low current density, the CO_2-to-CH_4 conversion is only 35% due to low conversion of steam and carbon

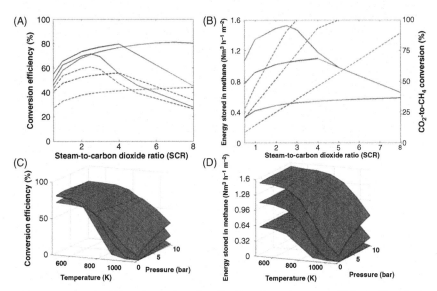

FIGURE 5.7   (A,B) Effect of steam-to-carbon dioxide ratio on the energy efficiency (A) and energy stored in methane (B); (C,D) effect of temperature and pressure on the energy efficiency (C) and methane production (D). *(From Ref. [14]. Copyright 2015 Elsevier Ltd.)*

dioxide in SOEC. At an applied current density of higher than 1 A cm^{-2}, the CO_2-to-CH_4 conversion almost reaches 100% at an SCR of larger than 5 due to superfluous hydrogen production and shortage of carbon dioxide. Consequently, the methane production rates at 1 and 1.5 A cm^{-2} are the same at SCR > 5, both decreasing with the rise of SCR. The optimal SCR is 7 at 0.5 A cm^{-2}, 4 at 1 A cm^{-2}, and 2.5 at 1.5 A cm^{-2}.

After determining the optimal SCR, the effects of the methanation temperature and system pressure on the electricity-to-methane efficiency and methane production rate at various current density and the corresponding optimal SCR are given in Fig. 5.7C and D. At high temperature, the methanation is not favored but its reversed reaction, that is, SMR, is promoted. As the temperature of the methanation reactor decreases to below 480°C, the electricity-to-methane efficiency exceeds to 70%. The operating pressure has a slight effect on the electrochemical performance of SOEC, while promote the methanation moving forward. The operating pressure has more apparent influence on the electricity-to-methane efficiency at higher methanation temperature and higher current density. But the influence of the operating pressure on the efficiency at low methanation temperature or low current density becomes much less. The less influence at low methanation temperature is attributed to enough CO_2-to-CH_4 conversion in the chemical equilibrium state, while the one at low current density is attributed to the lack of the reductive gas species such as H_2 and CO. The phenomenon is also revealed in the methane production rate curves in Fig. 5.7D. At a fixed SCR and current density, the methane production rate is closer to the maximum at lower temperature or higher operating pressure. Below 480°C, the methane production rate almost reaches the maximum no matter how much the operating pressure is. However, the methane production rate at a methanation temperature of 800°C and an applied current density of 1.5 A cm^{-2} significantly rises from 0.27 to $0.64 \text{ Nm}^3 \text{ h}^{-1} \text{ m}^{-2}$ as the operating pressure increases from atmosphere pressure to 1 MPa.

Consequently, a trade-off should be made between the electrical efficiency and the methane production rate. To realize a higher electricity-to-methane efficiency, the PtM system should operate at lower current density and higher SCR. However, to obtain a higher methane production and energy-to-methane efficiency, the PtM system is suggested to operate at higher current density and lower SCR.

5.2.3.3 Integrating the electrolyzer and methanation into one reactor

Previous discussions have offered two more feasible pathways, that is, the water electrolysis + methanation system and the H_2O/CO_2 co-electrolysis + methanation system. The other pathway is to converting H_2O/CO_2 directly to methane-rich product gas in one single reactor. When H_2O/CO_2 co-electrolysis is used for a PtM system, the SOEC reactor is capable of producing methane directly, especially at an elevated operating pressure.

The possibility of the direct PtM system based on one single SOEC reactor has been analyzed and discussed in detail in our previous study [8]. When an isothermal SOEC reactor is used for direct PtM, the produced gas at an H/C ratio of 10.5 and an applied current density of 0.6 A cm^{-2} is far from the allowed hydrogen content limitation of 15%. The hydrogen content in volume from the SOEC reactor at 550°C and 2.5 MPa is still as high as 56%, while the methane content in volume is only 43%. Despite the exergy efficiency reaches almost 70% (even higher at lower pressure and higher temperature), the energy stored in methane only accounts for at most 75% of the total stored energy. In other words, the energy-to-methane efficiency (in exergy) is only 50%, 20%–30% lower than the one in a PtM system integrating an extra methanation reactor. Jensen et al. [1] proposed a novel system integrating this isothermal SOEC reactor with subsurface storage of CO_2 and CH_4 (instead of the integration with the natural gas network). Therefore, this produced gas doesn't need to realize such a high CH_4 content requirement, and allows higher hydrogen content in the produced gas. In other words, the stored methane-rich gaseous fuels are the mixture of methane, hydrogen, and a bit of carbon monoxide. This system enables to realize large-scale energy storage with an overall round-trip efficiency of >70% (including BOP components) and an estimated storage cost of ~0.03 $ kWh^{-1}, comparable to pumped hydro storage and much more cost-efficient than other advanced energy storage technologies.

The temperature mismatch between SOEC and methanation leads to the limitation in the CH_4 content from the cathode outlet of an isothermal SOEC reactor. An SOEC reactor with a spatial temperature gradient as Fig. 5.8A shows is potential to alleviate this temperature mismatch. In this SOEC reactor, the reaction region can be divided into two zones: the SOEC zone and the methanation (MR) zone. Temperature in the SOEC zone distributes uniformly along with the gas flow direction (axial direction), while the one in the MR zone gradually decreases from the operating temperature of SOEC to the methanation temperature. Such an SOEC reactor can synchronously enhance the H_2O/CO_2 co-electrolysis and methanation. Through a suitable flow design, a better thermal integration can be realized. Assuming effective heat transfer of fuel and air streams, heat recovery from the outlet gas and adequate methanation activity in cathode, a direct PtM system using this nonisothermal SOEC reactor is potential to offer an exergy efficiency of higher than 76% at >600°C and <3 MPa, as in Fig. 5.8 shows. Compared with the two-step PtM system integrating H_2O/CO_2 co-electrolysis and methanation reactor, the direct PtM system performs more efficient below 23 bar. The optimal operating pressure was 8.15 bar, at which the H_2 volumetric fraction is exactly equal to 15%. At 8.15 bar, the direct PtM system can achieve an exergy efficiency of 81% when the SOEC zone is at 650°C and the MR zone varies from 650°C to 350°C, 3% higher than the two-step PtM system integrating H_2O/CO_2 co-electrolysis and methanation reactor and 4% higher than the SOEC-based PtM system integrating water electrolysis and methanation.

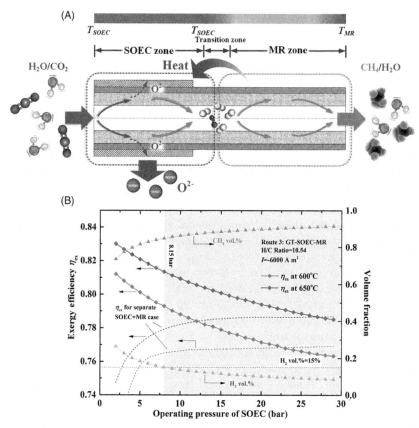

FIGURE 5.8 (A) Schematic of an SOEC reactor with a spatial temperature gradient and (B) exergy efficiency and outlet gas composition in the direct PtM system using the nonisothermal SOEC reactor at various pressure and temperature. *(From Ref. [8]. Copyright 2018 Elsevier Ltd.)*

5.3 Power-to-liquid systems

5.3.1 Power-to-methanol

Methanol is considered as an ideal liquid fuel due to its compatibility with fuel cells. PtMe can produce liquid methanol to realize easier transportation. The produced methanol can be transported to the end users, and converted to electricity using direct or indirect methanol fuel cells. Rivera-Tinoco et al. [15] integrated the PEMEC-based or SOEC-based water electrolysis subsystem with the methanol production process (CO_2 catalytic reduction to methanol) based on the simulation platform ASPEN Plus, respectively. Using this simulated systems, they evaluated the technological economy of the PtMe technology. In these two PtMe systems, the PEMEC stack operates at 15 bar, while the SOEC stack operates at ~850°C and 30 bar. The methanol production subsystems are the same in these two PtMe systems, as Fig. 5.9 shows. The methanol production subsystem consists of the

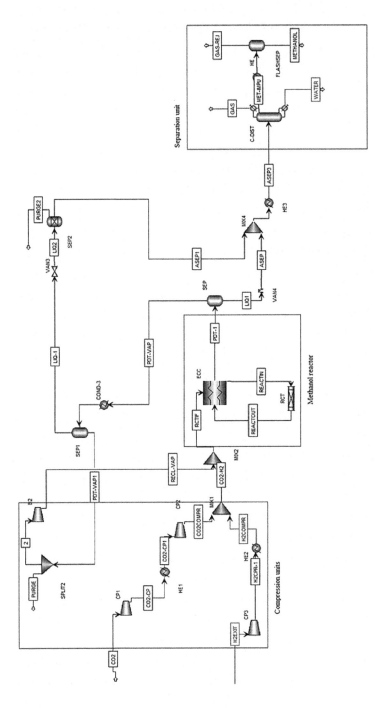

FIGURE 5.9 **System layout of the methanol production subsystem in the PtMe system.** *(From Ref. [15]. Copyright 2016 Elsevier Ltd.)*

compression unit, methanol reactor, separation unit, and other auxiliary equipment. The catalytic CO_2 reactor is fed with CO_2 and H_2 (from water electrolysis) and operates at 240–260°C and ~8 MPa using the $Cu/ZnO/Al_2O_3$ catalyst. The equilibrium CO_2-to-methanol conversion is only 46%, hence, the produced methanol with a purity of 99.9% is separated from the sequent distillation process. The simulation results reveal that the PEMEC-based PtMe system has an overall efficiency of 45%, while the overall efficiency of the SOEC-based PtMe system is ~10% higher than that of the PEMEC-based one [15].

In Rivera-Tinoco's study, they concentrated more on the economic analysis of both the PtMe systems. Considering the current technical maturity of both electrolyzers, they assumed that the lifespan of the PEMEC stack is 30,000 h with an investment cost of 600 kW^{-1} and that of the SOEC stack is only 8600 h with an investment cost of 5800 kW^{-1}. The design lifetime of the SOEC stack is only 29% of that of the PEMEC stack, while the SOEC stack is 8.6 folds more expensive than the PEMEC stack. Based on the assumption, they calculated the methanol costs to be 890 t^{-1} for the PEMEC-based PtMe system and 5460 t^{-1} for SOEC-based PtMe system [15]. Further, they analyzed the cost distributions and their sensitivity to the lifetime and cost of the electrolyzers. The results are shown in Fig. 5.10. Fig. 5.10A and B shows the cost distributions of both the PtMe systems for the reference case. For the first year, the replacement costs denote the initial CAPEX, and the CAPEX 0 ans is set to 0 in the first year. CAPEX both dominates the PtMe systems, accounting for 58% of the total annual costs for the PEMEC-based system and 93% for the SOEC-based system. However, OPEX of the SOEC-based system is only 7% of the total annual cost, much lower than the ratio in the PEMEC-based system due to extremely high investment cost of the SOEC stack. When we look on the whole lifetime (20 years), the total replacement cost of the SOEC-based system accounts for 78% of the total costs due to short lifetime of the SOEC stacks. The initial CAPEX of the SOEC-based system is only 12% of the total costs over the 20 years, and the total OPEX accounts for 10%. The majority of the initial CAPEX comes from the SOEC stack (98%), while that of the total OPEX comes from the consumed electricity (72%). As for the PEMEC-based system, the lifetime of the electrolyzers extends to more than 3 years, and the cost also remarkably reduces. Consequently, the replacement cost only accounts for 15% of the total costs, and the initial CAPEX for 10%. The majority of the total costs (75%) comes from the OPEX, in which the cost for electricity consumption accounts for 83%. As the electricity price increases from 20 to 150 MWh^{-1}, the total cost rises by 1500 t^{-1}. This increase can lead to the methanol cost for the PEMEC-based system rising almost 1.7 folds.

The cost distribution of the PEMEC-based and SOEC-based PtMe systems suggests different priorities for the cost reduction. The PEMEC-based system should put energy efficiency enhancement in the priority of the cost reduction, while the SOEC-based system should first reduce the investment cost of the SOEC stacks and lengthen their lifetime. This suggestion corresponds to the lifetime sensitivity demonstrated in Fig. 5.10C and D. The PEMEC stacks have

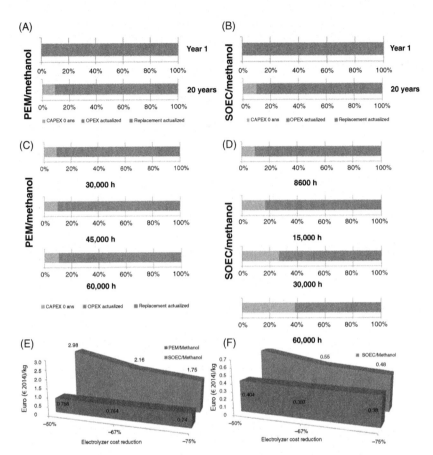

FIGURE 5.10   (A,B) Cost distributions of the PEMEC-based (A) and SOEC-based (B) PtMe systems in the first year and over the lifespan; (C,D) sensitivity of the lifespan to the cost distributions of the PEMEC-based (C) and SOEC-based (D) PtMe systems; (E,F) methanol production cost with the reduction in the electrolyzer cost in a reference case (E) and optimized case (F). *(From Ref. [15]. Copyright 2016 Elsevier Ltd.)*

enough lifetime and low investment cost. Thus, as the lifetime of the PEMEC stacks doubles, the total methanol cost only reduces by 4% and the ratio of the total replacement costs drops from 16% to 7%. However, as the lifetime of the SOEC stacks extends to 24,000 h, the total methanol cost can drop by 55% [15], and the replacement cost dramatically decreases.

Finally, Rivera-Tinoco evaluated the effect of the electrolyzer cost reduction on the methanol cost, as Fig. 5.10E and F shows [15]. In the reference case (Fig. 5.10E), a 75% reduction in the PEMEC cost only has a little effect on the methanol cost (from 860 t^{-1} by 17% to 740 t^{-1}), while a 75% saving of the SOEC cost significantly decreases the methanol cost by 68% from 5460 to 1750 t^{-1}. In an ideal case (Fig. 5.10F), they set the electricity price to 20 MWh^{-1}, and the lifetime of the PEMEC and SOEC to 90,000 and 60,000 h, respectively. The methanol cost in both the PtMe systems can reduce to below 700 t^{-1} with a 50% reduction of the electrolyzer cost. The methanol cost of the SOEC-based PtMe system becomes competitive to that of the PEMEC-based one. Particularly, the SOEC stack has a greater potential in the cost reduction due to no need for noble-metal electrodes.

5.3.2 Power-to-F-T liquid fuels

Fischer-Tropsch process (F-T) has been industrialized as a mature chemical engineering technology to produce liquid hydrocarbon fuels from syngas. H_2O/CO_2 co-electrolysis using SOEC can produce syngas with adjustable H_2/CO ratio, revealing a good compatibility with the conventional F-T synthesis.

Becker et al. [5] from Colorado School of Mines, USA carried out a comprehensive analysis on a hybrid system integrating H_2O/CO_2 co-electrolysis and F-T synthesis. Fig. 5.11A and B demonstrates the H_2O/CO_2 co-electrolysis

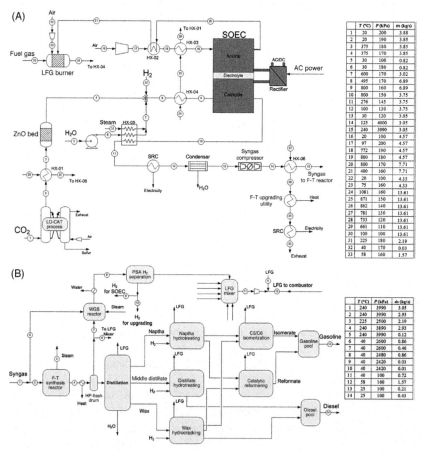

FIGURE 5.11 System layouts of the SOEC subsystem and F-T synthesis subsystem. *(From Ref. [5]. Copyright 2012 Elsevier Ltd.)*

subsystem and F-T synthesis subsystem In the H_2O/CO_2 co-electrolysis subsystem, CO_2 is first fed into a LO-CAT reactor and ZnO bed for removing the sulfur below 1 ppm. Then CO_2 mixes with steam and H_2, and exchanges heat with the outlet gas from the light fuel gas (LFG, C_{4-} hydrocarbons from the F-T synthesis) burner to 800°C. After heating the cathode gas, the outlet gas from the LFG burner further heats the air that has been preheated by the anode outlet gas. Therefore, the inlet gas from both anode and cathode are preheated to 800°C before fed into the SOEC stack. A part of the inlet H_2O feeding to the SOEC cathode comes from the steam produced from F-T synthesis and the water separated from the condenser. The inlet H_2O is preheated by the cathode outlet gas, and the remaining heat containing in the cathode outlet gas is used to generate electricity by using steam Rankine cycle (SRC). After condensation, the produced syngas is compressed to 40 bar and preheated 240°C for the sequent F-T synthesis. After preheating the SOEC subsystem and the produced syngas, the outlet gas from LFG burner can still keep at 781°C. This gas stream is further used to support heat for the sequent F-T synthesis, and then produce electricity through another SRC.

In the F-T synthesis subsystem, there are two main modules, that is, F-T synthesis reactor and the sequent F-T upgrading module. Becker et al. [5] uses the Anderson-Schulz-Flory (ASF) model to describe the complicated products from the F-T reactor operating at 240°C and 38–40 bar. The main products and their composition from the F-T reactor is shown in Fig. 5.12. The gaseous LFG with shorter carbon chains accounts for 5% in weight, which is mainly used for supplying heat or generating power. The C_5/C_6 hydrocarbons have a mass fraction of 7%, which can be blended into the gasoline after isomerization. Naphtha, denoting the C_{7-10} hydrocarbons, accounts for 16% of all the products in weight, which are the major components of the gasoline stock. The middle distillates, the C_{11-10} hydrocarbons, account for 34% of all the products

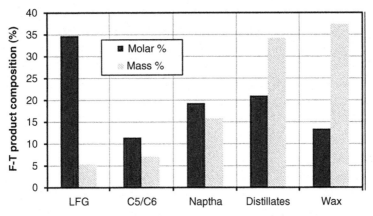

FIGURE 5.12 Products and their composition of F-T synthesis. *(From Ref. [5]. Copyright 2012 Elsevier Ltd.)*

in weight, which can be blended with the diesels. As the carbon number rises to higher than 20 (C_{20+}), the products belong to the wax, accounting for 37% of all the products in weight. The wax can be used to produce gasoline or diesel through the selective catalytic cracking. Typically, the conversion efficiency of the F-T reactor is 60%–90%, which is selected to 80% in Becker's study [5]. In other words, 80% of CO is converted into the hydrocarbon products in the F-T reactor. In the F-T synthesis subsystem, the gasoline and diesel are the target products. Thus, the products from the F-T reactor are further upgraded through naphtha hydrotreating, distillate hydrotreating, wax hydrocracking, C_5/C_6 isomerization, and catalytic reforming. The upgrading products from these processes are summarized in Table 5.1.

Becker et al. [5] analyzed the system efficiency using their built simulation platform. In the reference condition, the SOEC operates at 800°C and 1.6 bar, and the F-T reactor has a CO conversion of 80%. The SCR efficiency is set to 25%. The whole system (including the H_2O/CO_2 co-electrolysis subsystem and F-T synthesis subsystem) requires an electricity input of 54.5 MW_e, a CO_2 feedstock of 3.88 kg s^{-1} and a water supply of 181 gpm for generating 6.8 kgal gasoline and 12.7 kgal diesel per day. To evaluate the effectiveness of each part in this system, they defined a series of efficiency. The electricity-to-syngas efficiency represents the ratio of the energy stored in syngas (in higher heat value, HHV) to the total energy inputs to the SOEC subsystem (including the AC power input, energy in the cathode inlet gas and energy in the consumed LPG). The syngas-to-F-T liquids efficiency represents the ratio of the energy stored in the produced gasoline and diesel in HHV to the total energy inputs to the F-T synthesis subsystem (including the AC power input to the F-T subsystem and energy in the inlet syngas in HHV). Overall, the electricity-to-F-T liquids efficiency represents the ratio of the energy stored in the produced gasoline and diesel in HHV to the AC power input (to the whole system). The SOEC subsystem shows an electricity-to-syngas efficiency of 84.5%, the F-T subsystem reveals an syngas-to-F-T liquids efficiency of 49.4%. The whole

TABLE 5.1 Product composition in molar fraction [5].

Process	LFG	C_5/C_6	C_{7-10}	C_{11-19}	Isomerate	Reformate
C_5/C_6 isomerization	2.0%	0	0	0	98.0%	0
Naphtha hydrotreater	7.9%	14.2%	77.9%	0	0	0
Distillate hydrotreater	2.5%	0	0	97.5%	0	0
Wax hydrotreater	6.0%	5.0%	25.0%	65.0%	0	0
Catalytic reformer	14.2%	0	0	0	0	85.8%

TABLE 5.2 System efficiency with CO conversion in the F-T reactor [5].

Efficiency (%)	CO conversion in F-T reactor		
	70%	80%	90%
Electricity-to-syngas (HHV)	80.9	84.5	88.7
Syngas-to-F-T liquids (HHV)	43.0	49.4	55.9
Electricity-to-F-T liquids (LHV)	46.6	51.0	55.5
Electricity-to-F-T liquids (HHV)	50.1	54.8	59.7

system reveals the electricity-to-F-T liquids efficiency of 51.0% in lower heat value (LHV) and 54.8% in HHV. Table 5.2 shows these efficiencies at various CO conversion in the F-T reactor. As CO conversion drops to 70%, the syngas-to-F-T liquids efficiency reduces remarkably by 6.4 points of percentage, and the electricity-to-syngas efficiency also decreases by 3.6 points of percentage because of the reduction in the LFG. As a result, the electricity-to-F-T liquids efficiency (HHV) reduces by 4.7 points of percentage. On the contrary, a CO conversion of 90% significantly enhances these efficiency. The syngas-to-F-T liquids efficiency rises by 6.5 points of percentage, the electricity-to-syngas efficiency by 3.2 points of percentage, and the electricity-to-F-T liquids efficiency (HHV) by 4.9 points of percentage.

Fig. 5.13 shows the economics of the PtL system including the cost breakdown and cost sensitivity to the system parameters. The total investment cost for the PtL system is estimated to be 56.07 million dollars. Considering extra buildings and service facilities, the total direct cost needs an extra 12% cost, that is, 6.73 million dollars [5]. When it comes to the liquid fuels cost, the cost breakdown is shown in Fig. 5.13A. The liquid fuels including gasoline and diesel are converted into equivalent gasoline for comparability. The operating capacity factor (OCF) denotes the ratio of the time the PtL system operates at the maximum capacity to the lifetime. The intermittence of renewable power can significantly affect this factor. At a high OCF, the PtL system is utilized better, hence, the ratio of the capital cost to the total liquid fuels cost is much lower than that at a lower OCF. The ratio of the capital cost is only 29% at an OCF of 90%, and increases to 45% at an OCF of 40%. On the contrary, the costs for electricity consumption and CO_2 feedstock play a more important role in the total liquid fuels cost at a higher OCF. The electricity cost accounts for 54% at an OCF of 90%, and this ratio drops to 38% at an OCF of 40%. The costs for operation and maintenance (O&M) accounts for 13% of the total liquid fuels cost. O&M cost slightly drops at lower OCF due to the assumption that the SOEC replacement cost decreases linearly with OCF. The cost for CO_2 feedstock is only 4% of the total liquid fuels cost.

The sensitivities of the total liquid fuels cost to the electricity price, OCF, SOEC performance, and CO conversion in F-T reactor are estimated. Fig. 5.13B indicates that the liquid fuels cost can reduce down to $4.4 per gallon gasoline equivalent (GGE) with a low electricity price of $0.02 per kWh and an OCF of 90%, while soars up to $7.35 per GGE with an electricity price of $0.08 per kWh and an OCF of 40%. As the electricity price rises by $0.01 per kWh, the liquid fuels cost increases by ~$0.64 per GGE, that is, 15%–26%. Fig. 5.13C shows the effect of the SOEC performance on the liquid fuels cost. At a low OCF, the effect of the SOEC performance becomes more significant. As the ASR of the SOEC reduces from 1.5 to 0.25 Ω cm^2, the total liquid fuels cost reduces from $7.6 by 9% to $6.9 per GGE at OCF = 90%, while from $11.3 by 14% to $9.7 per GGE at OCF = 40%. When the PtL system is powered by a highly intermittent renewable power, a high-performance SOEC stack is more significant for the cost reduction. Fig. 5.13D shows the effect of the CO conversion in the F-T reactor on the total liquid fuels cost. At low electricity price, the syngas can be obtained with a lower cost, hence, the sensitivity of the total liquid fuels cost to the CO conversion in the F-T reactor is relatively low. As the electricity price rises, the cost for the syngas production also increases dramatically. Therefore, CO conversion in the F-T reactor has a more apparent influence on the total cost. At an electricity price of $0.10 per kWh, the liquid fuels cost rises by $1.7 per GGE (17%) to $8.6 per GGE with the CO conversion improving by 20%.

To sum up, it is vital for the economics of the PtL system to operate the PtL at its maximum capacity as much as possible (higher OCF). This is current challenge for the application of the PtL system driven by renewable power.

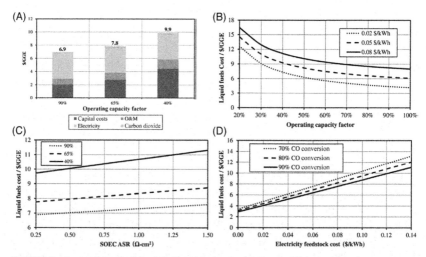

FIGURE 5.13 (A) Distribution of the liquid fuels cost; (B–D) sensitivity of the liquid fuels cost to electricity price (B,D), operating capacity factor (B,C), SOEC performance (C) and CO conversion in F-T reactor (D). *(From Ref. [5]. Copyright 2012 Elsevier Ltd.)*

Furthermore, Becker et al. [5] suggested that more incentives should be implemented to make the PtL technology more economically feasible.

5.4 Summary

In this chapter, we specifically showed some case studies of typical power-to-X systems, including PtH, power-to-syngas (PtS), PtM, PtMe, power-to-F-T fuels, etc. For each power-to-X pathway, we reviewed the system layouts, the technical characteristics, energy efficiency, and economic feasibility to reveal their potential in the future energy share and interaction.

References

[1] S.H. Jensen, C. Graves, M. Mogensen, C. Wendel, R. Braun, G. Hughes, et al. Large-scale electricity storage utilizing reversible solid oxide cells combined with underground storage of CO_2 and CH_4, Energy Environ Sci 8 (2015) 2471–2479.

[2] S. Becker, B.A. Frew, G.B. Andresen, T. Zeyer, S. Schramm, M. Greiner, et al. Features of a fully renewable US electricity system: optimized mixes of wind and solar PV and transmission grid extensions, Energy 72 (2014) 443–458.

[3] M. Beaudin, H. Zareipour, A. Schellenberglabe, W. Rosehart, Energy storage for mitigating the variability of renewable electricity sources: an updated review, Energy Sustain Dev 14 (2010) 302–314.

[4] Y. Luo, Y. Shi, S. Liao, C. Chen, Y. Zhan, C. Au, et al. Coupling ammonia catalytic decomposition and electrochemical oxidation for solid oxide fuel cells: a model based on elementary reaction kinetics, J Power Sources 423 (2019) 125–136.

[5] S.D. Ebbesen, C. Graves, M. Mogensen, Production of synthetic fuels by co-electrolysis of steam and carbon dioxide, Int J Green Energy 6 (2009) 646–660.

[6] B. Han, J. Mo, Z. Kang, F.Y. Zhang, Effects of membrane electrode assembly properties on two-phase transport and performance in proton exchange membrane electrolyzer cells, Electrochim Acta 188 (2016) 317–326.

[7] M.H. Miles, G. Kissel, P.W.T. Lu, S. Srinivasan, Effect of temperature on electrode kinetic parameters for hydrogen and oxygen evolution reactions on nickel electrodes in alkaline solutions, J Electrochem Soc 123 (1976) 332–336.

[8] Y. Luo, X. Wu, Y. Shi, A.F. Ghoniem, N. Cai, Exergy analysis of an integrated solid oxide electrolysis cell-methanation reactor for renewable energy storage, Appl Energy 215 (2018) 371–383.

[9] X. Zhang, C. Bauer, C.L. Mutel, K. Volkart, Life cycle assessment of power-to-gas: approaches, system variations and their environmental implications, Appl Energy 190 (2017) 326–338.

[10] Q. Fu, C. Mabilat, M. Zahid, A. Brisse, L. Gautier, Syngas production via high-temperature steam/CO_2 co-electrolysis: an economic assessment, Energy Environ Sci 3 (2010) 1382–1397.

[11] T. Schaaf, J. Grünig, M.R. Schuster, T. Rothenfluh, A. Orth, Methanation of CO_2—storage of renewable energy in a gas distribution system, Energy Sustain Soc 4 (2014) 1–14.

[12] Y. Luo, Y. Shi, N. Cai, Power-to-gas energy storage by reversible solid oxide cell for distributed renewable power systems, J Energ Eng 144 (2018) 4017079.

[13] M. Bailera, P. Lisbona, L.M. Romeo, S. Espatolero, Power to gas projects review: lab, pilot and demo plants for storing renewable energy and CO_2, Renew Sustain Energy Rev 69 (2017) 292–312.

[14] J.P. Stempien, M. Ni, Q. Sun, S.H. Chan, Production of sustainable methane from renewable energy and captured carbon dioxide with the use of solid oxide electrolyzer: a thermodynamic assessment, Energy 82 (2015) 714–721.

[15] R. Rivera-Tinoco, M. Farran, C. Bouallou, F. Auprêtre, S. Valentin, P. Millet, et al. Investigation of power-to-methanol processes coupling electrolytic hydrogen production and catalytic CO_2 reduction, Int J Hydrogen Energy 41 (2016) 4546–4559.

Chapter 6

Ammonia: a clean and efficient energy carrier for distributed hybrid system

Yu Luo[a], Yixiang Shi[b] and Ningsheng Cai[b]

[a]National Engineering Research Center of Chemical Fertilizer Catalyst (NERC-CFC), Fuzhou University, Fujian, China; [b]Department of Energy and Power Engineering, Tsinghua University, Beijing, China

6.1 Introduction

Renewable energy can be stored in the form of chemical energy using Power-to-X technology as Chapter 5 discussed, and the stored chemical energy can be efficiently converted into stable electricity as Chapter 4 discussed. The main options of the stored energy carrier X were also widely discussed in Chapter 5, typically involving H_2, syngas and hydrocarbon fuels. However, as a large amount of carbonaceous fuel-based fuel cells are used in the distributed energy systems (DES), carbon capture in the end-user DES is costly. Carbon-free energy carriers, represented by H_2, can realize zero CO_2 emission at the end-user side, which enables a more cost-efficient capture CO_2 from the large-scale fossil fuel-based power plants or chemical industries. In spite of high mass energy density of 142 MJ kg^{-1} (in higher heat value, HHV), H_2 has a low volumetric energy density [1]. At 25°C and 1 atm, volumetric energy density of H_2 is only 0.0108 MJ L^{-1}. As pressurized to 700 bar, H_2 has a volumetric energy density of 6 MJ L^{-1}. In other words, in comparison with gasoline, H_2 at 700 bar has a more than triple folds higher mass energy density, while only has less than 18% volumetric energy density [2]. Liquid H_2 can obtain a volumetric energy density of 10 MJ L^{-1}. Nevertheless, H_2 requires a temperature of lower than −240°C for liquefaction. Low volumetric energy density of H_2 results that H_2 storage without any efficient carriers requires extremely high pressure or extremely low temperature. H_2 is quite hard for efficient transport or storage. Consequently, a nonnegligible challenge regarding large-scale H_2 utilization in renewable energy storage and fuel cell systems is storage of H_2 [3]. To realize H_2 storage with high energy density by both weight and volume, the US Department of Energy (DOE) set targets of H_2 storage in 2020 as: Mass energy density of 11 MJ kg^{-1}

Hybrid Systems and Multi-energy Networks for the Future Energy Internet.
http://dx.doi.org/10.1016/B978-0-12-819184-2.00006-7

141

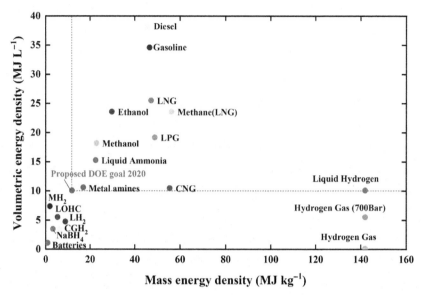

FIGURE 6.1 Energy density of various hydrogen carriers by weight and volume. *(Data from Ref. [2,4,5].)*

and volumetric energy density of 10 MJ L^{-1} [4,5]. To achieve these targets, a series of H$_2$ storage technologies have been developed involving cryocompression, metal hydrides, chemical hydrides, and adsorption materials [3].

Fig. 6.1 shows the energy densities of various alternative hydrogen carriers by weight and volume [2,4,5]. Some chemical hydrides such as methanol, liquid hydrocarbons, and liquid ammonia can achieve the targets of both mass energy density and volumetric energy density. Among these common chemical hydrides, only ammonia (NH$_3$) is carbon-free, hence, recently attracts much attention as a promising hydrogen carrier. Besides the carbon-free feature, NH$_3$ has a series of advantages [1]:

1. **High hydrogen content and energy density**: NH$_3$ contains 17.8% H in weight and 1 L liquid NH$_3$ can produce 108 g H$_2$ [6]. The H content of NH$_3$ outclasss the DOE targets for H$_2$ storage in 2015 [3, 7]: H content of 5.5 wt.% and 40 g H$_2$ L^{-1}. NH$_3$ has an energy density of 13.6 MJ L^{-1} by volume and 22.5 MJ kg^{-1} by weight [8,9], which are also higher than the DOE targets in 2020 mentioned earlier.
2. **Mature ammonia production technology**: NH$_3$ is one of the most common inorganic chemicals, widely used in the production of chemical fertilizer, plastics, fibers, explosives, and nitric acid, NO$_x$ removal, refrigeration, etc. More than 80% of NH$_3$ is used for the fertilizers, significantly enhancing global foot production [10]. Ammonia synthesis has a history of over one century [11]. With high demands in fertilizer and chemical industries, the global annual production of NH$_3$ is around 180 million tons [9]. Particularly

in China, the annual NH_3 production is around 57 million tons [12], accounting for 32% of the global NH_3 production. The use of ammonia as hydrogen carrier can fit in the current infrastructure and hence lower capital cost.

3. **Easy to store**: Above 8.6 bar at room temperature, NH_3 is in liquid form. Liquid ammonia is ~1.5 time that of liquid H_2 and 4 time that of the most advanced metal hydrides. In the real application, NH_3 is transported via metal tanks at 1.83 MPa so that NH_3 can remain in gaseous phase even above 50°C [12].

4. **Safe to store and use**: NH_3 is not flammable in air and has a much higher explosion limit (16%~25%) than gasoline vapor and natural gas [6]. Owing to its pungent odor, NH_3 detection is self-alarming. Just by smelling, it is possible to notice a leakage once NH_3 presence exceeds 5 ppm by volume. The Occupational Safety and Health Administration (OSHA) sets an acceptable exposure limit of 8 h at 25 ppm [4], while a NH_3 content of 5 ppm is well below any danger or damage. The partial pressure of NH_3 relative to toxicity is ~27 kPa, three orders of magnitude higher than those of methanol and gasoline vapor [6].

5. **No end-user CO_2 emission**: As mentioned earlier, NH_3 is a carbon-free hydrogen carrier. Despite more than 85% of the existing ammonia production comes from fossil fuels such as coal, oil, natural gas, etc [12], CO_2 emission from large-scale NH_3 production allows centralized CO_2 capture. The employment of carbon-containing hydrogen carriers inevitably leads to end-user CO_2 emission. The capture of distributed end-user CO_2 is not only much more expensive, but also is highly infeasible on board.

In this chapter, we will comprehensively discuss the concept and techno-economic feasibility of ammonia (NH_3)-based energy roadmap. The clean and efficiency technologies for NH_3 production and utilization will be reviewed, and the typical systems be compared with commercial ones. The technical maturity, existing challenges, and potential applications of these alternative NH_3-related technologies will be covered.

6.2 Ammonia-based energy roadmap

Based on the existing NH_3 application and potential NH_3-based energy conversion technologies, Fig. 6.2 shows the NH_3-related applications and energy roadmap. The most developed NH_3 synthesis is via Haber-Bosch process, that is, converting H_2 and N_2 catalytically into NH_3 at elevated temperature and pressure. To obtain high-purity H_2, two main approaches are optional: One is to convert water and carbonaceous fuels (fossil fuels like coal, natural gas and other hydrocarbons, or biomass) to produce H_2 via gasification or reforming process, water-gas shifting (WGS) process and CO_2 capture. The other one is to utilize renewable power (like wind power, solar power, hydropower, etc) to drive water electrolyzers. When driven by renewable power, NH_3 synthesis systems can realize carbon-free NH_3 production, and further form an absolutely

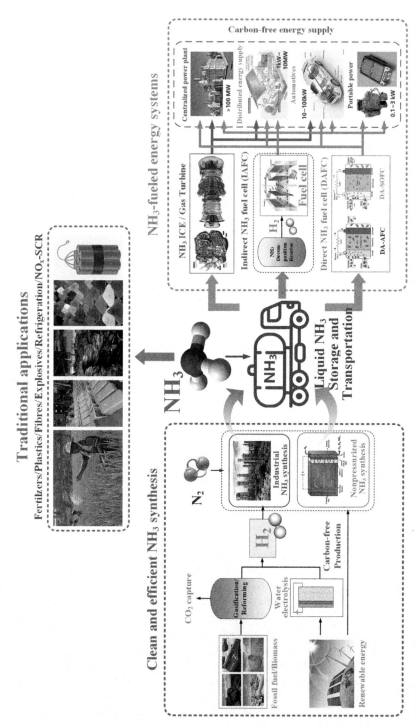

FIGURE 6.2 Conventional ammonia applications and ammonia-based energy roadmap.

carbon-free energy roadmap. Apart from Haber-Bosch process, some novel advanced nonpressurized NH_3 synthesis technologies via electrochemical, thermochemical, photochemical, and plasma catalytic approaches allow NH_3 synthesis in milder conditions at flexible scales, offering more potential options in the future NH_3-based energy roadmap.

Fig. 6.2 also presents the applications of NH_3 in the traditional fields and novel energy systems. As a fuel, NH_3 can generate stable power through internal combustion engines (ICE), gas turbines (GT) and fuel cells. NH_3-fueled ICEs and GTs are similar to the conventional ones, in which NH_3 is fired to generate high-temperature pressurized gas streams, then gas streams propel the rotors to convert thermal energy into mechanical energy, finally generates electricity using electric generators. NH_3 ICEs and GTs are applicable for automotives, DES, centralized energy systems (CES), etc. to supply electric power varying from 10 kW level to >100 MW. Differently, fuel cells can directly convert chemical energy in NH_3 into electric energy to offer electric power varying from 100 W level to >100 MW. Therefore, the potential applications include portable power, automotives, DES, CES, etc. According to the difference in reaction pathways, NH_3-fueled fuel cells can be classified into indirect NH_3 fuel cells (IAFC) and direct NH_3 fuel cells (DAFC). In an IAFC system, NH_3 first decomposes into H_2 and N_2 in an NH_3 decomposition reactor, then the produced H_2 fuels a fuel cell to supply continuous electric power. As for a DAFC, NH_3 is directly fed into a compact fuel cell device for power generation.

6.3 Current interest and projects on ammonia-based energy vector

6.3.1 Ammonia price

In United States, NH_3 production cost ($\sim$\$3.8/kg H) is approximately 28% higher than H_2 production ($\sim$\$3.0/kg H, industrial-scale purity) due to additional Haber-Bosch process [10]. In current industrial processes, the produced NH_3 can be easily liquefied to obtain high purity and low impurity content (CO, CO_2, sulfur, etc), however, to meet the requirement of fuel cells, the produced H_2 still needs extra purification processes to lower the contents of CO, sulfur, etc, for avoiding CO-poisoning and sulfur-poisoning. Considering this factor, the H_2 cost could be higher. Moreover, the end-user price should consider the cost of fuel transportation and storage. H_2 pipeline transport costs $\sim$\$1.87/kg H, while NH_3 pipeline transport only spends $\sim$\$0.19/kg H, about one tenth of H_2 pipeline transport. As for H storage, H_2 storage costs $\sim$\$1.97/kg H for a half-a-month storage period and $\sim$\$14.95/kg H for a semiannual period, but NH_3 storage only costs 3$\sim$4% of H_2 storage for the same storage period, i.e., $\sim$\$0.06/kg H for a half-a-month storage period and $\sim$\$0.54/kg H for a semiannual period [10]. In other words, after considering the costs of storage (15 days) and transportation, the end-user H_2 cost rises by over twofolds to higher than \$6.84/kg H, while the end-user NH_3 cost only rises by 6.6% to \$4.05/kg H, >41% lower than the end-user H_2 cost.

In Japan, the NH_3 price of cost, insurance and freight (CIF) was 170 yen/kg H ($1.58/kg H) in 2016, while the domestic liquid H_2 price was 1350~1800 yen/kg H ($12.6~16.7/kg H) and the price of compressurized H_2 was 1000~1100 yen/kg H ($9.3~10.2/kg H) in 2011~2015 [12,13]. The H price of NH_3 storage was lower than 17% of that of compressurized H_2 storage and 9~13% of that of liquid H_2 storage.

In China, the NH_3 cost (mainly based on coal as feedstock) varies from ¥2100/t ($297/t) to ¥2600/t ($367/t) [12,13]. Converted into equivalent H, the NH_3 price varies between ¥11.9/kg H ($1.68/kg H) and ¥14.7/kg H ($2.08/kg H). Industrial-level hydrogen production from coal costs ¥0.6~1.2/Nm^3 H_2, that is, ¥6.7~13.4/kg H ($0.95~1.89/kg H) [15]. To satisfy the requirement of fuel cells, the high-purity H_2 cost will rise to ~¥50/kg H (~$7.07/kg H). Considering H_2 compression, storage and transportation, the end-user H_2 price could be as high as ¥60~80/kg H ($8.48~11.3/kg H). In the end-user side, high-purity NH_3 price varies from ¥2550/t ($360/t) to ¥3260/t ($460/t) since January 2020, that is, ¥14.5~18.5/kg H ($2.05~2.61/kg H) [16]. Currently in China, the H price of NH_3 storage was 70~82% lower than that of compressurized H_2 storage.

6.3.2 Effectiveness of ammonia-based system

Zamfirescu and Dincer [5] presented a brief comparison among various fuel-fed vehicle systems in the system effectiveness, fuel cost per 100 km and driving range. The system effectiveness ε_r used by Zamfirescu and Dince also includes the refrigeration and other possibly recovered work (such as expansion work from the pressurized outlet gas). Among these 11 systems, 6 systems were fueled with NH_3, involving NH_3-fueled ICE, NH_3 decomposition reactor + ICE, NH_3 decomposition reactor + H_2/N_2 separation + ICE, DAFC, IAFC with H_2/N_2 separation and NH_3 electrolysis + PEMFC. When using ICE, the systems fueled with carbonaceous fuels only have effectiveness of lower than 28%, while that of NH_3-fueled ICE system reaches 44%. Even when integrating NH_3 decomposition, the system effectiveness was estimated to 31% after H_2/N_2 separation. As for the fuel cell-powered systems, the system effectiveness are much higher than those of the ICE-based systems. The FC-based system fueled with methanol only has an effectiveness of 33%, and the one powered by H_2 stored in metal hydrides has an effectiveness of 40%. In contrast, both the DAFC and IAFC systems have an effectiveness of higher than 44%, at least 4% higher than the metal hydride-based system. Due to the limitation in catalysts of NH_3 electrolysis at room temperature, the effectiveness of the NH_3 electrolysis + PEMFC system is only 20%. Except the NH_3 electrolysis + PEMFC system, the remaining NH_3-fueled systems have fuel costs of <2.4 $ per 100 km, at least 46% lower than the metal hydride-based system. As for driving range, gasoline has the longest driving range, that is, 825 km, due to the highest volumetric energy density. Most of the NH_3-fueled systems have driving ranges of higher than

TABLE 6.1 Performance of ammoina power systems and other systems [5].

Fuel/system	ε_r (%)	$100 km^{-1}	Range (km)
Gasoline/ICE	24	6.06	825
CNG/ ICE	28	6.84	292
LPG/ ICE	28	5.10	531
Methanol/reforming + FC	33	9.22	376
H_2 metal hydrides/FC	40	4.40	142
NH_3/direct ICE	44	1.57	592
NH_3/Th decomp, ICE	28	2.38	380
NH_3/Th decomp Sep, ICE	31	2.15	420
NH_3/direct FC	44	1.52	597
NH_3/Th. decomp + Sep, FC	46	1.45	624
NH_3/electrolysis	20	3.33	271

380 km, and the driving ranges of the NH_3-fueled ICE, DAFC, and IAFC systems can reach as far as ~600 km, >fourfolds longer than the metal hydride-based fuel cell system. Table 6.1

6.3.3 Ammonia-based energy projects

Currently, ammonia-based energy vector has attracted the attention of a number of developed countries, including Japan, United States, United Kingdom, Australia, etc [17–21]. The Cross-Ministerial Strategic Innovation Program (SIP) of Japan has promoted the establishment of "Green Ammonia" consortium consisting of 19 companies and 3 research institutes including Tokyo Gas, IHI Corporation, the Japan Science and Technology Agency (JST), JGC Corporation, Osaka Gas, Chugoku Electric Power, Chubu Electric Power, AIST, Nippon Shokubai Co., Ltd., etc [22]. Currently, they have launched a series of projects to combine NH_3 with current energy systems. Such as, IHI Corporation and Tohoku University were developing a gas turbine fueled with NH_3/CH_4-rich fuel, and Chugoku Electric Power was trying to co-fire NH_3 and coal in their established power plants [9]. A remarkable demonstration carried out by JGC Corporation and the National Institute of Advanced Industrial Science and Technology (AIST) is the world's first NH_3-based hybrid system with a complete NH_3-based energy storage cycle involving renewable power-to-NH_3 and NH_3-to-power [19]. In collaboration with Nippon Shokubai Co., Ltd., Toyota Industries Corporation, Mitsui Chemicals Inc. and IHI Corporation, Eguchi's group from Kyoto University demonstrated a 1000 h operation of a 1 kW-class DA-SOFC stack.

In United States, the Advanced Research Project Agency-Energy (ARPA-E) from the Department of Energy (DOE), United States [23] launched the Renewable Energy to Fuels through Utilization of Energy-dense Liquids (REFUEL) program to develop scalable technologies for converting renewable energy into chemical energy stored in energy-dense liquid fuels and subsequent converting liquid fuels into H_2 or electricity for transportation or distributed energy supply. In REFUEL program, 32.7 million dollars were granted to support 16 projects, 81% of which selected NH_3 as the ideal CNLF. In Europe, a cooperation "Green Ammonia" project supported by UK Science and Technology Funding Council (STFC) was carried out by Siemens, Oxford University, Cardiff University, and Rutherford Appleton Laboratory. This project established a renewable power-to-NH_3 system with a daily yield of 30 kg in Rutherford Appleton Laboratory and has been in operation since June 2018 [18]. Supported by the Fuel Cells and Hydrogen Joint Undertaking (FCH JU), Alkammonia Project managed by a consortium led by AFC Energy integrated aims to develop high-efficiency and cost-effect IA-AFC and DA-AFC to supply power for Base-Transceiver Stations (BTS) in remote areas [20]. Recently, a €10 million funding offered from FCH JU supported another NH_3-powered fuel cell project called "ShipFC project" run by 14 European companies and institutions led by the Norvegian cluster organization NCE Maritime CleanTech [24]. In this project, the scale of IAFC will be upgraded from 100 kW-level to 2 MW-level to power a ship solely for 3000 h per year. Viking Energy is in charge of the installation of the IAFC and expected to finish the IAFC system at the end of 2023. If this project finishes the demonstration successfully, this IAFC system could be the largest NH_3-powered fuel cell system over the world.

In Australia, renewable power-to-NH_3 technology has been a nationwide focus. Most renewable H_2 projects are being run by NH_3 producers according to Rystad Energy's data in April 2020 [25]. Currently, Australian government is trying to support the development of H_2 projects enabling to their industrial scale-up in the coming 5 years. According to Rystad Energy's statistic data [25], the national water electrolysis projects in Australia have reached more than 700 MW, 58% of them are attributed to NH_3 producers for carbon-free NH_3 synthesis. In these NH_3 producers, the Yara Pilbara, the second biggest NH_3 producer, has the largest capacity, reaching up to ~240 GW (solar power-to-NH_3). Another report from Ammonia Energy Association [26] indicates that the Yara Pilbara project could climb up to higher than 1 GW by 2030, and another project, that is, The H2U Port Lincoln pilot, could help to form a series of "Hydrogen Hubs" with a potential electrolysis capacity of 3 GW.

6.4 Hybrid systems for ammonia production

As estimated, more than 96% of global NH_3 production systems (APS) use Haber-Bosch process which has been discovered by two Nobel Laureates Haber and Bosch for about one century to produce ~97% of global nitrogen fertilizers [9,27,28]. Large NH_3 demand and stable $N\equiv N$ bond lead to a large amount of energy consumption for NH_3 production, accounting for approximately 1.8~3.0% of global energy consumption [9,27]. As mentioned above, H_2 used in Haber-Bosch process is mainly produced from fossil fuels, therefore, inevitably leading to high CO_2 emission during NH_3 production. Without effective CO_2 capture, current CO_2 emission from NH_3 production reaches approximately 1.2% of the worldwide CO_2 emission caused by human activities. Current industrial Haber-Bosch process requires catalysts, and operates at high temperature (400~600°C) and high pressure (10~25 MPa). To realize a clean and efficient NH_3 production, current Haber-Bosch process needs optimization. On the one hand, lowering operating temperature and pressure of Haber-Bosch process can enhance the system efficiency and economics, and also allow the development of novel distributed systems for more flexible NH_3 production. On the other hand, carbon-free energy sources like renewable energy, nuclear energy, etc, can replace existing fossil fuels as the energy sources to produce green ammonia.

6.4.1   System schematic and flow charts

Nowadays, an APS using Haber-Bosch process typically includes two subsystems: H_2 production subsystem and Haber-Bosch subsystem. To clarify the energy efficiency and economic feasibility, Smith et al. [28] compares two typical NH_3 production routes based on Haber-Bosch process, i.e., one CH_4-fueled route and one electrically driven route. The former one represents the conventional highly carbon-intensive NH_3 production using fossil fuels because natural gas supports half of fossil fuel-fed NH_3 production [28], and the latter one the carbon-free NH_3 production driven by hydropower, wind power, solar power, nuclear power, etc. Fig. 6.3 shows the flow charts of these two typical NH_3 production routes [28]. Fig. 6.3A shows that in the H_2 production subsystem, CH_4 is first reformed at 3~25 bar into syngas-rich gas stream via primary and secondary steam methane reforming (SMR) reactors, then syngas-rich gas stream is upgraded to H_2-rich gas stream via one- or two-stage WGS process and purified via CO_2 removal and methanation. Here, primary SMR reactor operates at 850~900°C with external heat input from CH_4 combustion, and secondary SMR reactor operates at 900~1000°C. In the secondary SMR reactor, the autothermal reaction processes are realized by the addition of air to provide heat released from catalytic partial oxidation. The remaining N_2 from the air can be used for the subsequent Haber-Bosch process, and high-temperature steam produced during SMR also has high additional value as a by-product. Considering NH_3 synthesis catalysts could suffer from CO poisoning, CO content should be controlled below an acceptable value. Therefore, WGS reactor can convert H_2O and most of CO into CO_2 and H_2. Limited by the WGS equilibrium, there is still excess CO remaining unreacted. The final methanation process further helps to reduce CO content to an acceptable level, but produces CH_4 as an inert

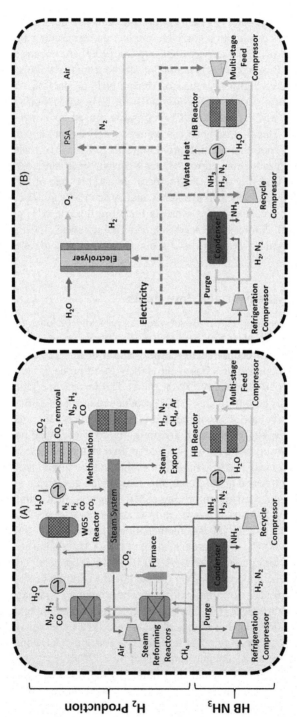

FIGURE 6.3   System flow charts of ammonia production: Haber-Bosch process coupling with a conventional methane-fueled hydrogen production subsystem (A) and a water electrolysis subsystem. *(From Ref. [28], Copyright The Royal Society of Chemistry 2020.)*

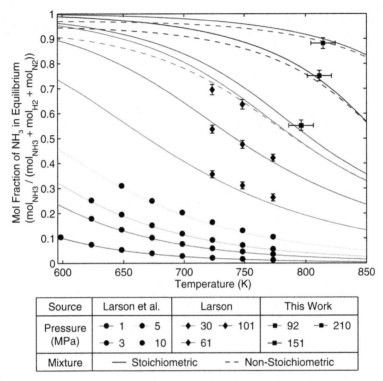

Source	Larson et al.		Larson		This Work	
Pressure (MPa)	● 1	● 5	◆ 30	◆ 101	■ 92	■ 210
	● 3	● 10	◆ 61		■ 151	
Mixture	— Stoichiometric			- - Non-Stoichiometric		

FIGURE 6.4   Equilibrium ammonia molar fraction as a function of temperature and pressure. *(From Ref. [29], Copyright 2015 Elsevier.)*

gas fed and accumulated in the Haber-Bosch subsystem. In the CH_4-fueled system, the high-temperature gas stream from SMR outlet and the cooling stream from WGS can be recovered to heat high-pressure steam, and further used to drive steam turbines for compression. In the whole system, the largest energy consumption is attributed to two main compressors, that is, one is used for the air compression of secondary SMR inlet, and the other one for the compression of feeding gas before Haber-Bosch reactor.

In the Haber-Bosch subsystem, Haber-Bosch reactor is the key component for NH_3 production. An elevated temperature or pressure can promote the kinetics of NH_3 synthesis. In thermodynamics, the equilibrium curve of NH_3 synthesis in Fig. 6.4 shows the rise in operating pressure is thermodynamically favored, but the rise in operating temperature unfavorable. Fe-based catalysts are conventional catalysts for Haber-Bosch process, which generally prefers to operate at 400~500°C and 100~250 bar. Limited by the catalytic activity of Fe-based catalysts, conventional Haber-Bosch reactor has to operate at a relatively high temperature to obtain acceptable reaction kinetics. As a result, the single-pass equilibrium conversion of NH_3 synthesis is relatively low, that is, 10%~15%. Industrial solution is to recycle the unreacted gas after condensation for NH_3 separation. Multiply gas cycles can lead to the accumulation of CH_4-rich inert gas, hence, the accumulated inert gas is purged and used for SMR heating [28]. To enhance the system efficiency and NH_3 production, Haber-Bosch reactor can use novel NH_3 synthesis catalysts with higher catalytic activity, in which Ru-based catalysts are suitable options. Compared with a conventional Fe-based catalyst, Ru-based catalyst can promote the single-pass conversion of NH_3 synthesis from 8% to 15% at 430°C and 100 bar [30], potentially significantly reducing both operating temperature and pressure of Haber-Bosch process. British Petroleum (BP) has developed a commercial Ru/C catalyst. Using the catalyst developed by BP, Kellogg Advanced Ammonia Process (KAAP) successfully retrofitted the world's first large-scale NH_3 synthesis plant using Ru-based catalyst in 1992 [31,32]. So far, KAAP process has retrofitted 7 world-scale NH_3 synthesis plants with a total capacity of 2000 t NH_3 per day [31].

In spite of the majority of global fossil fuel-based NH_3 production using natural gas as feedstock, coal-fed NH_3 production accounts for more than 70% of total NH_3 production in China [12]. In comparison with the CH_4-fed APS, a coal-fed APS uses a coal gasifier to replace SMR reactors. Generally, a coal gasifier operates at higher than 700°C and lower than 100 bar. Because of large oxygen demand for coal gasification, air separation units (ASU) is required, and the separated N_2 can be directly used for Haber-Bosch process. National Engineering Research Center of Chemical Fertilizer Catalyst (NERC-CFC) from Fuzhou University has developed a commercial Ru-based catalyst [33–38]. In 2018, [a]Protection and New Materials Co., Ltd. (BSEPNM) successfully applied the self-developed Ru-based catalyst in conjunction with conventional Fe-based catalyst to one 10,000 t/year-scale coal-based NH_3 synthesis plant located in Suqian, Jiangsu, China. The NH_3 synthesis plant operates Haber-Bosch reactor at $\sim$100 bar and lower than 440°C with an H/N ratio of 2.6~2.8 to obtain a single-pass conversion of $\sim$18% [39]. One year later, NERC-CFC and BSEPNM further scaled the novel coal-based NH_3 synthesis plant up to a capacity of 200,000 t/year [40]. As estimated, the NH_3 synthesis plant using the Ru-based catalyst can improve NH_3 production by more than 30%, and reduce energy consumption by more than 10% [41]. Considering a high-pressure coal gasifier (at $\sim$80 bar) generally operates at a higher operating pressure than SMR reactor, the pressure of the main gas streams between H_2 production subsystem and Haber-Bosch subsystem become close. Besides, high-activity Ru-based catalysts show the potential to reduce the operating pressure of Haber-Bosch reactor. As the operating pressure of Haber-Bosch reactor reduces to lower than 80 bar, operating pressure of SMR reactor matches with that of Haber-Bosch reactor. Therefore, the compressor before the Haber-Bosch reactor can be removed to realize a compacter and more flexible system integration. In addition, biomass is also a carbon-rich fuel. The biomass-fueled APS has a similar

---

a. NERC-CFC cooperating with Beijing Sanju Environmental.

system configuration to the coal-fueled APS [42], hence, these progress in coal-fueled NH_3 production can help to realize efficient and flexible biomass-to-NH_3.

With a trend of chemical industry electrification, conventional centralized chemical plants will gradually transition to distributed chemical plants with flexible scales. A low-pressure Haber-Bosch process increases the flexibility of NH_3 production plants, leading to a better adaptability with the APSs driven by carbon-free electric power like renewable power and nuclear power. Fig. 6.3B shows the flow chart of an electricity-driven APS, in which electricity is the only external energy input instead of conventional fossil fuels. Here, the major change in APS is the H_2 production subsystem. In an electricity-driven APS, water electrolysis is used to produce high-purity H_2, thus, the only need for H_2 purification is drying for H_2O removal. For the other feedstock of Haber-Bosch process, the acquisition of high-purity N_2 could vary with the scale of NH_3 production. In a large-scale NH_3 production plant, N_2 can be obtained from cryogenic distillation, which has been used for ASU in coal-fueled APSs. At a smaller scale, pressure swing adsorption (PSA) should be more economical technology. In addition, membrane separations could also be an alternative technology relying on its technical maturity and produced N_2 purity. Differently from fossil fuel-fueled APSs, electricity-driven APSs need no oxygen, but produce a large amount of oxygen from water electrolysis and air separation. Thus, pure oxygen is a major byproduct in the electricity-driven system.

6.4.2 Energy efficiency and economic analysis

An early design for an electricity-driven APS should date to 1970s, when Grundt et al. [43] from Norsk Hydro in Norway integrated Haber-Bosch process with alkaline electrolysis cells (AEC) driven by hydropower for a small-scale application. According to Grundt's calculation [43], an electricity-driven APS consumed 34.1×10^6 BTU energy to produce 1 t NH_3, slightly lower than that of a natural gas-fueled system (34.5×10^6 BTU/NH_3) and 17% lower than a coal-fueled system (41×10^6 BTU/NH_3). They also performed a brief economic analysis to reveal that capital charges and electricity price are two main factors. Some notable numbers they provided showed that 1 t NH_3 production consumes 10 MWh electricity, that is, each \$0.01/kWh to be paid for will increase the total cost by \$100. In recent years, with an increasing need for sustainable and clean energy sources, potential energy penalties related to CO_2 arises new attention to carbon-free NH_3 production. Therefore, current NH_3 production technologies should be evaluated using a comprehensive index combining energy efficiency, economic feasibility and CO_2 emission instead of only using economic feasibility. Smith et al. [28] performed a re-evaluation of a CH_4-fueled APS and an electricity-driven one using current new perspective. Considering the environmental factor, they estimated that an optimized and high-efficiency CH_4-fueled APS causes ~ 1.6 t CO_2 equivalent per tonne NH_3 production ($t_{CO2\text{-}eq} \, t_{NH3}^{-1}$), while one renewable power-driven one only emits $0.12 \sim 0.53$ $t_{CO2\text{-}eq} \, t_{NH3}^{-1}$, more than 67% lower than an CH_4-fueled one. They concluded that in an CH_4-fueled APS, H_2 production via SMR accounts for $\sim76\%$ of CO_2 emission, that is, approximately 1.22 $t_{CO2\text{-}eq}$ t_{NH3}^{-1}, and the remaining CO_2 emission (~0.38 $t_{CO2\text{-}eq}$ t_{NH3}^{-1}) is attributed to CH_4 combustion used for heating and steam production. In a typical wind power-driven APS, major CO_2 emission comes from power generation from wind turbine ($0.35\sim0.49$ $t_{CO2\text{-}eq}$ t_{NH3}^{-1}, up to capacity and windspeed), accounting for $\sim92\%$ of total CO_2 emission. The remaining CO_2 emission is attributed to N_2 separation and gas compression. In comparison, an APS driven by grid power emits ~1.65 $t_{CO2\text{-}eq}$ t_{NH3}^{-1} based on current electric grid status, even higher than CH_4-fueled APS.

Smith et al. [28] also summarized the energy consumption of various APSs for the past decades and their potential minimum. The lower heat value (LHV) of NH_3 is only 18.6 GJ t_{NH3}^{-1}, revealing the chemical energy stored in NH_3. Actually, energy losses are inevitable, resulting in an energy consumption exceeding 18.6 GJ t_{NH3}^{-1}. Befor 1970s, APSs mainly used coal as feedstock. Early in 1950s, the coal-fed APSs were still not well optimized and required more than 60 GJ t_{NH3}^{-1}, revealing energy losses of more than 40 GJ t_{NH3}^{-1} and an efficiency of only $\sim30\%$. After one decade, the energy consumption of the coal-fed APSs can reduce to lower than 40 GJ t_{NH3}^{-1} and improve system efficiency by more than 16%. After 1970s, natural gas gradually replaced the role of coal in APSs. Natural gas-fed APSs required to consume $30\sim37$ GJ t_{NH3}^{-1} in early stage, and were currently optimized to $27\sim32$ GJ t_{NH3}^{-1}. As Smith's evaluation, a current electricity-driven APS approximately consumes 38.2 GJ t_{NH3}^{-1} in the assumption of 60% electrolysis efficiency, consisting of 35.5 GJ t_{NH3}^{-1} for H_2 production and ~2.7 GJ t_{NH3}^{-1} for N_2 separation. Theoretically, the minimum energy requirement of a CH_4-fed APS is 22.2 GJ t_{NH3}^{-1}, consisting of 17.7 GJ t_{NH3}^{-1} CH_4 feedstock (in LHV) and 4.5 GJ t_{NH3}^{-1} heat requirement. In an electricity-driven APS, the reduced high-temperature heat requirement potentially enhances thermal integration, thus, the minimum energy requirement further decreases to 21.3 GJ t_{NH3}^{-1}. Another future alternative APS, that is, direct electrochemical NH_3 synthesis from water and nitrogen, can further reduce the minimum energy requirement to 19.9 GJ t_{NH3}^{-1} at room temperature and atmosphere pressure [28]. Direct electrochemical NH_3 synthesis at room temperature was first reported by Pickett and Tlarmin in 1985 [10,44]. However, current direct electrochemical NH_3 synthesis is still far away from commercialization, hence, unable to obtain acceptable NH_3 selectivity and yield. Meanwhile, differently from current gas-phase Haber-Bosch process, direct electrochemical NH_3 synthesis requires to separate NH_3 from the product solution, which is also a costly and energy-consuming process. To sum up, current technical maturity of direct electrochemical NH_3 synthesis even leads to energy consumption much higher than any current APSs in use or being commercialized. Based on energy consumptions of these APSs mentioned earlier, Smith et al. further offered the details in energy loss distributions of these APSs as shown in Fig. 6.5 [28]. For the CH_4-fed APS, a series of improvements

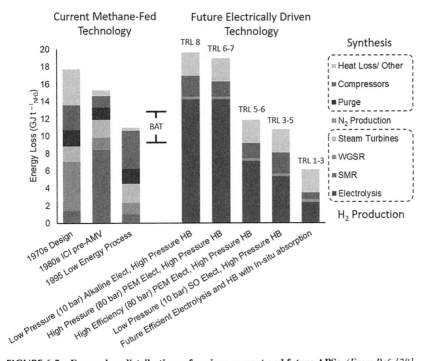

FIGURE 6.5 Energy loss distributions of various current and future APSs. *(From Ref. [28], Copyright The Royal Society of Chemistry 2020.)*

has been made to save energy consumption since 1970s, including novel technology development like Selexol process for efficient CO_2 removal, enhanced catalysts and reactors enabling better activities and higher SMR pressure, more efficient and compacter system integration, etc. [28]. In present CH_4-fed APS, total energy losses of ~ 11 GJ t_{NH3}^{-1} (corresponding to a system efficiency of 63%) are mainly attributed to compressors ($\sim 36\%$), H_2 production ($\sim 32\%$, including SMR, WGS, and H_2 purification) and steam turbines ($\sim 18\%$). As for electricity-driven APSs, despite energy saving in H_2 purification and compression, an AEC-driven APS performs less efficiently (49%) than a well-developed CH_4-fed APS due to low electrolysis efficiency (60%). The electrical energy loss used for AEC reaches ~ 14 GJ t_{NH3}^{-1}, $\sim 74\%$ of total energy losses in an AEC-driven APS. Enhanced electrolysis efficiency can exceed by more than 15% to higher than 75% by using proton exchange membrane electrolysis cells (PEMEC) and solid oxide electrolysis cells (SOEC) in the future. The use of efficiency electrolyzers can reduce the electrical energy loss caused by electrolysis by more than a half to less than 7 GJ t_{NH3}^{-1}, leading to a competitive system efficiency (61% for PEMFC at 80 bar and 63% for SOEC at 10 bar) to well-developed CH_4-fed APS. Due to easier liquid compression than gas compression and faster electrochemical kinetics at high pressure, a pressurized aqueous electrolyzers (AEC and PEMEC) can save ~0.8 GJ t_{NH3}^{-1} compression-related energy losses than those at atmosphere pressure or a high-temperature steam-fed SOEC [28]. Electric compressor performs at a higher efficiency than steam turbine-driven compressor to reduce 3.5 GJ t_{NH3}^{-1} energy losses. However, little need for high-temperature steam or other heat sources leads to extra heat losses of 2.6 GJ t_{NH3}^{-1}. This part of heat losses is potential to supply nearby heat demands such as district domestic heating, steam supply for chemical plants, greenhouse food plants, etc. As electrolysis efficiency further improves to ~90% and low-pressure Haber-Bosch process is commercialized to operate at ~3 bar by integrating with in situ NH_3 absorption, the future electricity-driven APS potentially reduces energy losses to ~6 GJ t_{NH3}^{-1}, involving only ~2.5 GJ t_{NH3}^{-1} eletrolysis-related energy loss and ~0.5 GJ t_{NH3}^{-1} compression-related one.

Considering the energy efficiency and scale flexibility are the major requirements for the future renewable power-driven APSs, the major challenges are efficient low-pressure Haber-Bosch process and flexible and energy-saving NH_3 separation. Focusing on these two challenges, Smith et al. [28] compares the energy losses and capital costs of the Haber-Bosch subsystems using various alternative system layouts at different-level operating pressure in detail as Fig. 6.6 reveals. In this comparison study, Haber-Bosch reactors at three operating pressure levels (1.5~3 bar, 20 bar and 150 bar) matched with three currently available or future NH_3 separation technologies, respectively, and the Haber-Bosch subsystem in current CH_4-fed APS was compared as a base case.

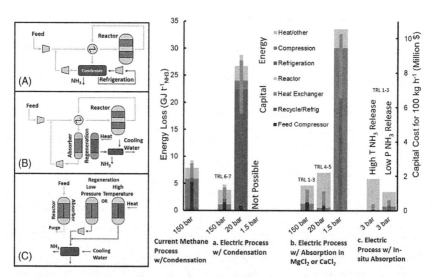

FIGURE 6.6 Energy losses and capital costs of the Haber-Bosch subsystems at different-level operating pressure (1.5~150 bar) using various alternative NH_3 separation technologies. (A) Condensation; (B) absorption in $MgCl_2/CaCl_2$; (C) in situ absorption. *(From Ref. [28], Copyright The Royal Society of Chemistry 2020.)*

The Haber-Bosch subsystem of a CH_4-fed APS with $100 \, kg \, h^{-1} \, NH_3$ yield results in $\sim 7.6 \, GJ \, t_{NH3}^{-1}$ energy losses and requires $\sim \$2.5$ million capital costs. The majority of energy losses is attributed to compression, and most of capital costs are used for gas compressors (\$1.7 million) and heat exchangers (\$0.7 million). In contrast, energy losses and capital costs can be reduced to $\sim 3 \, GJ \, t_{NH3}^{-1}$ and approximately by using the electricity-driven APS at 150 bar, mainly attributed to significant efficiency improvement caused by replacing steam turbine-driven compressor with electric one. Current large-scale high-pressure Haber-Bosch process typically uses a condenser at $-25°C \sim -33°C$ and ~ 140 bar to separate NH_3 from the outlet of Haber-Bosch reactor, and the remaining outlet gas is recycled back to mix with feeding gas as Fig. 6.6A shows. As operating pressure of Haber-Bosch process drops, the equilibrium NH_3 yield significantly also decreases despite high-activity catalysts such as Ru-based catalysts could make atmosphere-pressure Haber-Bosch process applicable in the future. Lower NH_3 content leads to increasing difficulty in NH_3 separation using condensation. A pressure ranging from 20 bar to 30 bar, the condenser requires a lower temperature ($\sim -33°C$) for NH_3 separation. However, low single-pass NH_3 yield remarkably increase energy consumed for refrigeration and capital costs related to refrigeration and heat exchangers, leading to both total energy losses and capital costs triple higher than those of current CH_4-fed APS. As pressure further drops to near 1 atm, the condensation temperature should be lower than an acceptable temperature ($< -45°C$) [28]. Absorption in crystalline salt-based absorbents such as $MgCl_2$ or $CuCl_2$ exhibits high capability in NH_3 separation even at low NH_3 partial pressure (~ 0.002 bar) [45]. The electricity-driven Haber-Bosch subsystem above 20 bar using this absorption technology has lower total energy losses and capital costs than current CH_4-fed one, while that at 1.5 bar has quite low single-pass equilibrium conversion (below 1%), leading to extremely high energy consumption and capital costs used for vast recycles and heat exchange. Compared with condensation-based Haber-Bosch loop, absorption-based one leads to higher heat losses due to the regeneration temperature of higher than 300°C [28]. To reveal the potential of the future atmosphere-pressure Haber-Bosch process, Smith et al. [28] proposed a novel Haber-Bosch reactor integrating in situ absorption operating at ~ 3 bar, in which absorbents can flexibly operate in pressure-swing mode or temperature-swing mode. In the former mode, the capital costs related to compressor are higher but energy losses are low, while in the latter mode, high heat losses are inevitable due to frequent temperature variation but capital costs are quite low. No matter which mode is selected, the estimated capital costs is much lower than any other compared Haber-Bosch subsystems, and the total energy losses are also competitive in the premise of enough technical maturity.[b]

---

b. As Smith stated, the advance in H_2 production and Haber-Bosch process will make a second NH_3 revolution: Transition from fossil fuels to renewable power and from large-scale centralized plants to flexible-scale distributed systems.

Zhang et al. [42] carried out another techno-economic study on green APSs at a capacity of 50,000 t/year, including CH_4-fed APS, biomass-fed APS and electricity-driven APS. They concluded that an electricity-driven APS has a system efficiency of more than 74%, higher than that of a CH_4-fed APS (61%) or a biomass-fed APS (44%). Zhang's study [42] revealed a similar efficiency of a CH_4-fed APS to Smith's calculation (63%) [28], while at least 11% higher efficiency of an electricity-driven APS using SOEC than Smith's corresponding results (63% for an SOEC-based electricity-driven APS). The difference could be attributed to the different system integration strategies. High-temperature outlet gas from SOEC can be used to produce steam or support the heat demand of Haber-Bosch reactor, thus, potentially realizes a more efficient thermal integration. Zhang et al. [42] varied a series of operating parameters such as operating temperature, pressure, flow rates, etc, and drew one efficiency-cost curve for each APS as Fig. 6.7A shows. The CH_4-fed APS has the lowest NH_3 production cost of $374~387 t^{-1} and the shortest payback time of 4.6~4.9 years among these three APSs, and the electricity-driven one has the highest cost of $544~666 t^{-1} and the longest payback time of 6.5~10.5 years. These three APSs all reveal a negative correlation between system efficiency and NH_3 production cost. According to the efficiency-cost curves, they defined two typical operating conditions as maximum-efficiency point (MEP) and minimum-cost point (MCP), respectively. Fig. 6.7B,C give the details of investment costs, operating costs and incomes for three APSs operating at MCP, respectively. The majority of investment costs in a CH_4- or biomass-fed APS are attributed to heat exchangers and Haber-Bosch process, corresponding to Smith's study [28]. In addition, extra costs of gasifier and ASU remarkably increase the investment costs of a biomass-fed APS. In the electricity-driven APS at MCP, besides the investment costs for Haber-Bosch process, high NH_3 production cost is mainly attributed to two factors, that is, expensive SOEC stacks and high electricity price. As each SOEC stack price drops to $470, NH_3 production cost reduces by 17% to $453 t^{-1} and payback time shortens to 5.3 years, which are competitive to those in a CH_4-fed APS. With electricity price decreases from $73 $MW^{-1}h^{-1}$ to $35 $MW^{-1}h^{-1}$, NH_3 production cost can save by 48% to $281 t^{-1} and payback time is only 4.4 years, exceeding those of a CH_4-fed APS. Fig. 6.7C indicates that even at an electricity price of $73 $MW^{-1}h^{-1}$, total electricity costs is still lower than the incomes from NH_3 selling. Once pure oxygen can be properly used, O_2 selling will also be a significant income of the electricity-driven APS. When using surplus renewable power or that cannot be used by grid, particularly in China, electricity price could remarkably reduce to promote high-efficiency, economical, and green NH_3 synthesis.

6.5 Ammonia-fueled hybrid systems

Despite NH_3 synthesis has developed for over one century, using NH_3 as an energy carrier is still a new hot topic with rising attention of low-carbon energy vectors. Supported by the Cross-Ministerial Strategic Innovation Program (SIP)

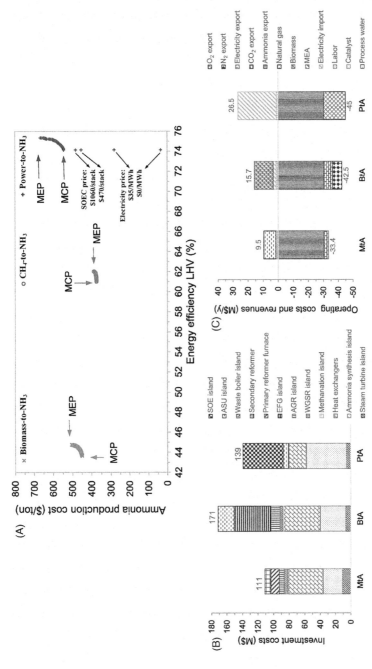

FIGURE 6.7 Comparison of three APSs in (A) relation between NH₃ production cost and system efficiency in various operating conditions; (B) investment cost distribution in minimum-cost operating condition; (C) operating cost and Income distribution in minimum-cost operating condition. (*Modified from Ref. [42], Copyright 2019 Elsevier.*)

of Japan, the "Green Ammonia" consortium led by Tokyo Gas was organized for seeking the demonstration of NH_3 as a hydrogen carrier [9,22]. In REFUEL program launched by ARPA-E from DOE 81% of funded projects (13 of all the 16 projects) selected NH_3 as the target liquid fuels. To realize this target, the technologies of NH_3-to-H_2 and NH_3-to-power are significant bases of the future NH_3-based energy roadmap.

6.5.1 Ammonia-fueled engines

A conventional pathway of NH_3 utilization is through well-developed heat engines such as ICE and GT based on NH_3 combustion. Early-stage innovation of NH_3-to-power has been proposed as an alternative fuel of steam engines used for small locomotive in the first industrial revolution [9]. NH_3 can enhance combustion and alleviate the adverse effects like knocking owing to its high octane number, while also introduce new issues related to NH_3 corrosion so that brass or copper is not applicable in NH_3-fueled engines [9,46]. Considering low flame speed and high resistance to auto-ignition of NH_3, NH_3-blended bi-fuels are first considered as a practical approach to realize NH_3 combustion in a heat engine, involving gasoline/NH_3, diesel/NH_3, H_2/NH_3, CH_3OH/NH_3, C_2H_5OH/NH_3, NH_4NO_3/NH_3, etc [9]. Korean Institute for Energy Research (KIER) developed a 30% gasoline/70% NH_3-fueled car named the AmVeh to reduce at least 70% CO_2. As they estimated, using the AmVeh in 20% of vehicles can reduce 10 million annual CO_2 emissions in Korea, that is, 20% of Korean 2020 target of CO_2 emission reduction [21]. In power plants, Kansai Electric Power cooperating with five other companies has begun to use NH_3 and coal for power generation via co-firing, which also potentially reduces CO_2 emission from coal plants by more than 20% and further leads to 40 million ($\sim$3%) annual CO_2 emission in Japan with the assumption of 70 coal plants switching to NH_3/coal co-firing [47]. Despite this technology could increase electricity price by $\sim$30% to 7 yen $kW^{-1}h^{-1}$, this price is still quite competitive to that of nuclear power plants ($\sim$10 yen $kW^{-1}h^{-1}$) and natural gas-fueled power plants ($\sim$14 yen $kW^{-1}h^{-1}$) in Japan.

H_2/NH_3 mixture is a specific bi-fuel of all because H_2 is potentially produced from NH_3 decomposition. Therefore, the ICE or GT can be only fueled with NH_3 by integrating with an NH_3 decomposition reactor. One experimental study have shown that down to 5 vol.% H_2-blended NH_3 is able to offer quite ideal power response [48], and another study revealed 10 vol.% H_2-blended NH_3 had the optimal H_2/NH_3 ratio regarding to efficiency and power output [49]. Ezzat and Dincer [50] proposed two 118 kW NH_3-fueled vehicle systems integrating NH_3 decomposition reactor as Fig. 6.8A/B show. One of them only uses ICE, and the other one integrates ICE and PEMFC. The thermodynamic assessment revealed that the addition of NH_3 decomposition reactor can recover heat from exhaust gas of ICE to avoid more than 13% exergy destruction. The maximum exergy destruction was found in the ICE. Thus, the former ICE-based

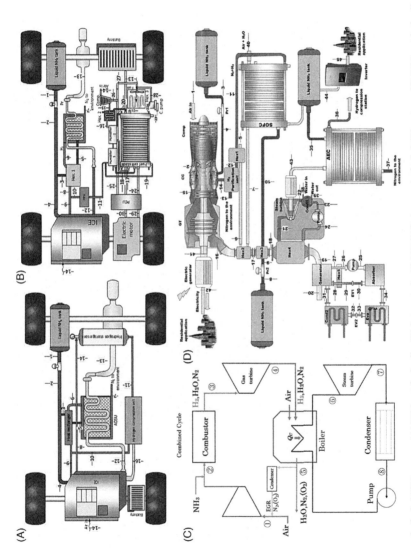

FIGURE 6.8 Ammonia-fueled ICE system: (A) Without fuel cells; (B) Integrated with PEMFC; Ammonia-fueled GT system: (C) Without fuel cells; (D) Integrated with SOFC. *(Part (A)/(B) from Ref. [50], Copyright 2017 Hydrogen Energy Publications, (C) from Ref. [51], Copyright 2020 Elsevier and (D) from Ref. [52], Copyright 2019 Elsevier.)*

system only has an estimated efficiency of 35% (First Law) and 38% (Second Law), while the integration of PEMFC in the latter system enhances system efficiency by 27% (First Law) and 25% (Second Law) [50]. In addition, Keller et al. [51] analyzed an NH_3-fueled combined-cycle GT-steam turbine (ST) system (as Fig. 6.8(c) shown) in thermodynamics. A high-pressure ratio and low exhausted gas recirculation ratio (EGR) can enhance system efficiency. Keller et al. also observed that there is a tradeoff between efficiency and NO_x formation from combustor. Considering a viable turbine inlet temperature of $< 2000°C$, an EGR of 0.6 was suggested to obtain an optimal efficiency of $\sim 61\%$ at a pressure ratio of 20 and an equivalence ratio of 1. Ezzat and Dincer [52] proposed a novel NH_3-fueled SOFC-GT hybrid system as Fig. 6.8D shows, and evaluated the energy and exergy efficiency of this system. To utilize the exhaust heat of GT, Rankine cycle and an absorption chiller were also integrated. The electricity produced from Rankine cycle drove an NH_3 electrolysis cells (AEC) for H_2 production. The absorption chiller was used for industrial refrigeration. The main part of this system is the integration of GT and SOFC, used to supply the majority of electricity. Totally, this hybrid system exhibits a system efficiency of 59% (First Law) or 51% (Second Law) with the capability of co-generating 17.5 MW net electricity (4.5 MW from SOFC and 13.0 from GT), 109 kg h^{-1} H_2, 36 t h^{-1} hot water (80°C) and 0.36 MW cold.

Compared with NH_3-fueled ICE, NH_3-fueled GT has relatively low technical maturity. Supported by Innovate UK, Siemens cooperating with Oxford University, Cardiff University and Science and Technology Facilities Council [17] launched "Green Ammonia" project to demonstrate an NH_3 energy storage system integrating renewable power-driven NH_3 synthesis and NH_3-to-power at Rutherford Appleton Laboratory. This system has the capability up to produce 30 kg NH_3 per day [18]. Using the produced NH_3, Siemens and their cooperators planned to generate stable electricity via NH_3-fueled ICEs in current stage, via NH_3-fueled GTs in 2023 and via novel NH_3 fuel cells in the future. In the same year, JGC Corporation and AIST successfully ran the world's first NH_3-based hybrid system consisting of 20 kg/day renewable power-driven NH_3 synthesis subsystem and 47 kW NH_3-fueled GT-based power generation subsystem [19]. This system first achieved a complete and scale-up energy chain from power-to-NH_3 to NH_3-to-power.

6.5.2 Ammonia-to-hydrogen

As a hydrogen carrier with high H mass fraction, H_2 production from NH_3 (NH_3-to-H_2, AtH) via NH_3 decomposition is one significant application of NH_3 in the NH_3-based energy roadmap. AtH enables to use NH_3 in current H_2-fueled power generation devices like ICE, GT, fuel cells, etc. Apart from enhancing thermal efficiency of heat engine-based DHS [50], AtH is also an essential component of IAFCs. Conventionally, AtH has been used in the semiconductor industry, metallurgical industry and other related industries as an economical reducing shielding gas. After H_2/N_2 separation, AtH can produce high-purity H_2. Due to easy storage and high volumetric H capacity of NH_3, *on-site* H_2 production from NH_3 can significantly reduce the investment cost of H_2 refueling infrastructures. Various catalysts like Ni-based and Ru-based catalysts have been widely studied to reveal high NH_3 decomposition catalytic activity. Currently, a series of Ni-based catalysts have been commercialized; however, require to operate above 800°C to achieve enough high NH_3 conversion. To improve flexibility, safety and energy efficiency of AtH and enhance compatibility with subsequent H_2-fueled generation devices, NH_3 decomposition should operate at a lower temperature. In thermodynamics, equilibrium NH_3 conversion exceeds 99% above 400°C and at 1 bar. NH_3 decomposition is the reverse reaction of industrial NH_3 synthesis reaction, that is, Haber-Bosch reaction. As high-activity NH_3 synthesis catalysts, Ru-based catalysts also reveal high NH_3 decomposition activity, hence; potentially lower the required temperature of AtH system. In our lab, an industrial-level Ru/C catalyst has been developed to achieve NH_3 conversion of higher than 96% at 400°C and a gaseous hourly space velocity (GHSV) of 15,000 mL g_{cat}^{-1} h^{-1}. After 500 h lifetime test, the Ru/C catalyst revealed no apparent degradation. Due to endothermic reaction, NH_3 decomposition requires heat input to support continuous reaction. A simple approach has been used in the industries is electrical heating, however, electricity-to-heat leads to a decrease of energy grade, that is, lower system efficiency, from the perspective of Second Law of Thermodynamics. Researchers are spending their efforts on designing novel NH_3 decomposition reactors with highly thermal integration. Kim et al. [53] designed a microsystem integrating one Ru-based NH_3 decomposition micro-reactor and one micro-combustor for burning H_2/NH_3/air mixture, in which the microsystem can produce 20 mL min^{-1} H_2 with an NH_3 conversion of 97% but emit 158 ppm NO_x in the optimized operating condition.

To match with current H_2-related industries, the product gas from NH_3 decomposition (mainly 75% H_2/25% N_2 and unreacted NH_3) needs to be purified to obtain high-purity H_2 despite N_2 is generally an inert component. Therefore, a complete AtH system needs a series of purification processes involving NH_3 removal and H_2/N_2 separation as Fig. 6.9A. For example, to supply H_2 applicable for fuel cells, an H_2 refueling station should enhance H_2 purity to higher than 99.97% (i.e., a total impurity content of less than 300 ppm) and lower NH_3 content to less than 0.1 ppm according to ISO14687-2 [54]. There are several alternative options, that is, adsorption [55], absorption [56], and selective catalytic oxidation (SCO) of NH_3 [57], and H-permeable membrane [58], for trace NH_3 removal. For H_2/N_2 separation, cryogenic separation, membrane separation, and PSA are optional. In mediate- to small-scale H_2 production, cryogenic separation is not cost-effective due to large equipment investment and high-energy consumption. Membrane and PSA are the main consideration for the application in H_2 refueling stations. In contrast, PSA can obtain high-purity H_2 with relatively low H_2 recovery and high investment costs, while commercial

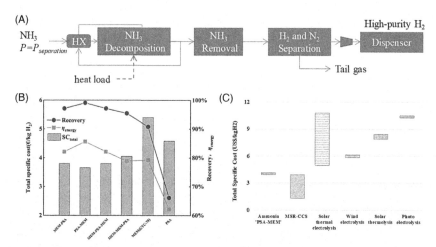

FIGURE 6.9 (A) A typical ammonia-to-hydrogen system schematic; (B) Comparison among various H_2/N_2 separation configurations in recovery, efficiency and specific costs; (C) Comparison in total specific cost among various H_2 production technologies. *(From Ref. [59], Copyright The Royal Society of Chemistry 2020.)*

polymeric membranes are hard to obtain acceptable H_2 purity in spite of high recovery and low investment costs. To realize a feasible and economical AtH at a scale suitable for H_2 refueling stations, our group [59] carried out a systematic techno-economic analysis on a 280 Nm^3 h^{-1} H_2 refueling station and comprehensively compared a series of H_2/N_2 separation configuration based on current commercial H_2/N_2 separation technologies, that is, PSA and membranes. In this study, we validated our developed PSA and membrane models with experimental data. Aiming to obtain H_2 with a purity of >99.95%, we proposed 8 different H_2/N_2 separation configurations and calculated the corresponding energy efficiency and total specific costs for comparison and optimization as shown in Fig. 6.9B and Table 6.2. Results revealed that H_2 refueling station integrating *on-site* AtH and "PSA-to-membrane" separation subsystem is optimal in both efficiency and specific costs with the highest H_2 recovery of >95%. The optimized H_2 recovery is >29% higher than the one using PSA-only separation. Owing to high H_2 recovery, system efficiency increases from 62% by 24% to 86%, and total specific cost reduces by 20% to 3.65 €/kg_{H2}. Fig. 6.9C revealed that the optimized total specific cost of AtH is comparable to H_2 production from SMR-based system with carbon capture and storage (CCS), and >28% lower than other carbon-free H_2 production technologies. According to ISO14687-2, commercial polymeric membrane is hard to obtain enough high H_2 purity. Pd membrane is a novel mediate-temperature membrane technology (at 300~550°C) applicable for H_2 ultra-purification (up to 99.999999%) [60]. Therefore, Table 6.2 reveals that Pd membrane-based H_2/N_2 separation can offer an energy efficiency of ~80%. However, Pd membranes are still at the early stage of commercialization with relatively high price, leading to the total

TABLE 6.2 Efficiency and costs of ammonia-to-hydrogen systems using various H_2/N_2 separation configurations [59].

Separation configuration		η_{energy}	Total specific cost (€/kg$_{H2}$)
PSA		40%~62%	4.58–7.42
Pd membrane		79%~81%	4.67–6.68
Membrane-to-PSA		82%	3.81
PSA-to-Membrane		86%	3.65
MEM-PSA-MEM		82%	3.81
MEM-MEM-PSA		79%	4.05
MEM-PSA-MEM-RR		83%	3.80–3.83
MEM-MEM-PSA-RR		77~81%	4.02–4.33

specific cost of 4.67~6.68 €/kg$_{H2}$. Due to temperature match with NH_3 decomposition, Pd membranes are potential to integrate with NH_3 catalytic decomposition into a compact reactor, enabling direct production of high-purity H_2 in one single NH_3-fed reactor [61].

6.5.3 Indirect ammonia fuel cells

In IAFC, NH_3 decomposition reactor and fuel cells are integrated into one system, in which H_2/N_2 mixture produced from NH_3 decomposition can be directly fed into fuel cells, hence, H_2/N_2 separation can be removed. Various fuel cell technologies have different preferred temperature and fuel compatibility, resulting that system layouts could be more or less different. Supported by the Fuel Cells and Hydrogen Joint Undertaking (FCH JU), Alkammonia Project managed by a consortium led by AFC Energy integrated aims to develop high-efficiency and cost-effect NH_3-fueled AFC to supply power for Base-Transceiver Stations (BTS) in remote areas [20]. Based on a 3 kW Alkammonia DHS shown in Fig. 6.10A, Cox et al. [62] performed life cycle assessment and estimation of levelized cost of electricity (LCOE). Due to high NH_3 tolerance of AFC, the product gas from NH_3 decomposition reactor can be directly fed into AFC without NH_3 removal after heat recycle. Air is supplied to AFC by a blower, and a CO_2 scrubber is needed for CO_2 removal before air enters AFC. Fig. 6.10B shows the environmental impact and non-renewable primary energy consumption of each system components per kWh power generation in the Alkammonia system fueled with NH_3 from natural gas. Considering a 10-year operation as

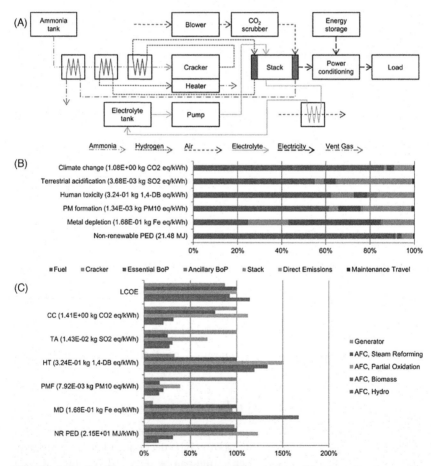

FIGURE 6.10 (A) Alkammonia system diagram; (B) environmental impact and non-renewable primary energy consumption of each system components; (C) LCOE, environmental impact and non-renewable primary energy consumption of diesel generator-powered system and AFC-powered system fueled with ammonia from various energy sources. *(From Ref. [62], Copyright 2014 Elsevier.)*

well as recycling and disposal of materials, results showed that in the whole lifecycle, each kWh electricity production from the 3 kW Alkammonia system was estimated to cause a series of energy or environmental impacts, involving climate change (1.08 kg equivalent CO_2 emission), terrestrial acidification (3.68×10^{-3} kg equivalent SO_2 emission), human toxicity (0.324 kg equivalent 1,4-DB), particulate matter (PM) formation (1.34×10^{-3} kg equivalent PM10) and metal depletion (1.68×10^{-1} kg equivalent iron), as well as non-renewable primary energy consumption of 21.48 MJ [62]. Except metal depletion, NH_3 consumption accounted for the majority of environmental or energy impacts. Therefore, green NH_3 production is the essence of environmental impact reduction. Fig. 6.10C further compared the LCOE, environmental impact and non-renewable primary energy consumption of diesel generator (DG)-powered system and AFC-powered system fueled with NH_3 from various energy sources. Fig. 6.10C indicates that electricity cost of the AFC system using NH_3 produced from natural gas was 14% high than that of the diesel engine-driven system despite climate change, terrestrial acidification, and PM formation were much lower than those of the diesel engine-driven system. Cox's calculation revealed that the diesel engine system costed more in fuel, while the AFC system using NH_3 costed twice more in hardware than the diesel engine system. Even the cost for hardware accounted for almost 60% of the total cost in the AFC system, revealing the economic feasibility can be remarkably improved by reducing hardware costs. When comparing the AFC system from various NH_3 sources, Cox et al. [62] found that the AFC system using NH_3 from biomass has the lowest LCOE, while that using NH_3 from hydropower has the highest one regarding to energy source costs. AFC systems using NH_3 from natural gas and heavy hydrocarbon feedstock have similar LCOE, while the latter has significantly more serious environmental impacts except metal depletion due to lower energy efficiency of heavy hydrocarbon-to-H_2. Biomass and hydropower are both renewable energy sources. The AFC systems using NH_3 from these two renewable energy sources can dramatically reduce at least a half of equivalent CO_2 emission and more than 70% non-renewable primary energy consumption. Hydropower generators require much more metal materials, thus, hydropower-driven NH_3 production leads to $> 50\%$ more metal depletion than other NH_3 production. Finally, Cox et al. [62] performed a sensitivity analysis based on the NH_3-fueled AFC system, revealing that LCOE is mainly affected by fuel cost, power density, lifetime and system efficiency, while environmental impacts are dominated by system efficiency.

Differently from AFC, PEMFC can be significantly affected by NH_3-containing reacting gas (even ppm-level NH_3). The poisoning of NH_3 could be attributed to the conductivity degradation of polymer membrane/electrode caused by increased NH_4^+ concentration, lower anode catalyst activity due to NH_3 competitive adsorption or lower cathode activity caused by NH_3 crossover [63]. ISO14687-2 requires that NH_3 content should be no more than 0.1 ppm [54], which is impossible to be realized only by NH_3 decomposition due to the

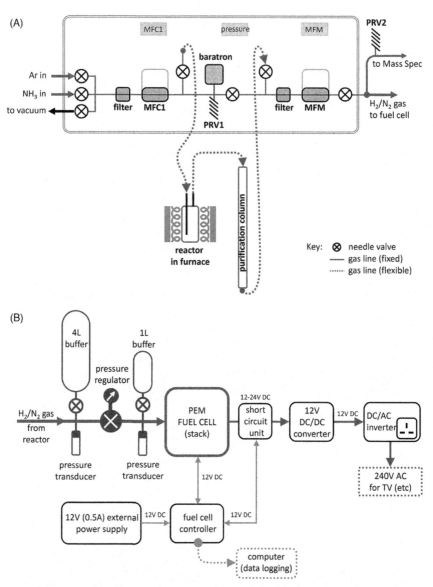

FIGURE 6.11 100 W indirect ammonia PEMFC systems in Rutherford Appleton Laboratory including H₂ production subsystem (A) and PEMFC subsystem (B). *(From Ref. [56], Copyright 2016 Elsevier.)*

equilibrium limitation. Hunter et al. [56] developed an indirect NH₃ PEMFC (IA-PEMFC) system consisting of an NH₃ decomposition reactor, a post-reactor gas purification column and a 100 W-scale PEMFC as shown in Fig. 6.11A,B. In the NH₃ decomposition reactor, light metal imides, and amides were used as the catalyst and the heat demand from NH₃ decomposition was supplied by an electric furnace. The post-reactor gas purification column with $MgCl_2$ absorbent was used to remove the undecomposed NH_3 below 0.1 ppm. The purified mixture of 75% H_2 + 25% N_2 is fed into a 100 W PEMFC to power an LED television.

Fig. 6.11B shows before the H_2/N_2 mixture gas was supplied to PEMFC, a 4 L buffer and a 1 L buffer were placed in the upstream position and the downstream position of the pressure regulator, respectively, to minimize the impacts of the upstream gas flow on PEMFC and ensure that the inlet fuel gas of PEMFC at 1.5 bar. A 12 V DC/DC converter and a DC/AC inverter was linked in series to power a digital LED television. A linear purge strategy was also designed by Hunter to ensure the continuous operation of the IA-PEMFC system at various power output.

Cha et al. [55] from Korea Institute of Science and Technology (KIST) demonstrated a 1 kW-class IA-PEMFC system as Fig. 6.12A shows. This system mainly consisted of a liquid NH_3 tank, an NH_3 decomposition reactor, balance of plants (BOP, including heat exchanger, burner, and etc.), an NH_3 adsorption column, and a 1 kW-class PEMFC. Differently from Hunter's system, this system uses $Ru/La-Al_2O_3$ as the catalyst and the NH_3 decomposition reactor was integrated with an iso-butane-fueled burner. Besides, various NH_3 adsorbents were evaluated for limiting NH_3 content in the inlet fuel stream of PEMFC, including $CaCl_2/Al_2O_3$, $MgCl_2/Al_2O_3$, HZSM-5, HY and a commercial 13X zeolite. Comparison revealed that commercial 13X zeolite had the highest NH_3 adsorption capacity of ~ 3 mmol$_{NH3}$ g$_{ad}^{-1}$ [55]. The compact NH_3 decomposition reactor (shown in Fig. 6.12B) was well designed with alternating trapezoidal-shaped channels welded into a cone shape around the burner to enhance heat exchange between NH_3 decomposition channels and heat flow fluid. The neighboring NH_3 decomposition channels had opposite flow directions. Iso-butane was used to preheat the NH_3 decomposition reactor to 550°C. After stable operation, heat source can be alternatively replaced by excess H_2 recirculated from PEMFC. A heat exchanger was compactly placed above the NH_3 decomposition reactor for preheat inlet NH_3 stream. The IA-PEMFC system from KIST has been stably operated for over 2 h using both exhausted H_2 and exhausted H_2 + iso-butane, respectively. Experiments revealed that as the applied current of PEMFC increased, NH_3-to-H_2 efficiency and system efficiency both increased. Using the recirculated H_2 as the heat sources, the NH_3-to-H_2 efficiency can climb up to $\sim 85\%$ and the overall system efficiency increased to $\sim 50\%$. Cha et al. [55] used the IA-PEMFC system to power a drone. Compared with only 14 min flight duration using battery, the flight duration using IA-PEMFC with 3.4 L NH_3 tank can extend the flight duration to ~ 4.1 h. The compact IA-PEMFC was able to produce 52 kWh energy with 4.9 wt.% when using light aluminum tank. Fig. 6.13C estimated that the system energy densities and hydrogen storage capacities as a function of total system weight. At low power output, the energy density is quite low due to the weight of essential component materials. As system weight exceeds 50 kg, the gravimetric energy density remarkably

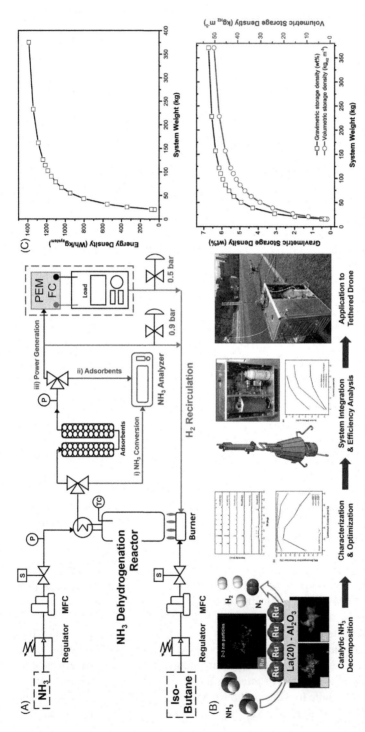

FIGURE 6.12 1 kW-level indirect ammonia PEMFC systems. (A) System schematic; (B) its realization from catalysts to demonstration, and (C) system energy densities and hydrogen storage capacities with total system weight. *(From Ref. [55]. Copyright 2018 Elsevier.)*

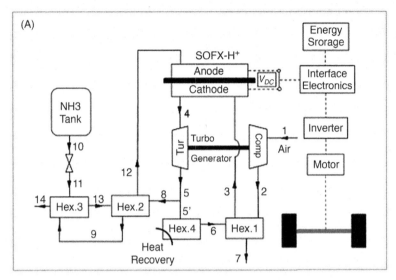

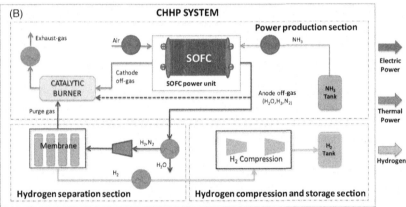

FIGURE 6.13 System schematic. (A) Distributed hybrid system integrating DA-SOFC and gas turbine; (B) combined hydrogen, heat and power system using DA-SOFC. *(From Ref. [71], Copyright 2011 Hydrogen Energy Publication, LLC; and Ref. [72], Copyright 2018 Elsevier.)*

rises to higher than 800 Wh kg^{-1}. According to the estimation of hydrogen storage capacities, the designed IA-PEMFC system from KIST is possible to satisfy the US DOE target in 2020 (i.e., 4.5 wt% and 30 g_{H2} L^{-1}) when system weight exceeds 50 kg. As system further scales up to ≥21 kg, the energy density of the IA-PEMFC DHS was expected to surpass current state-of-art Li-ion batteries (~350 Wh kg^{-1}) [55]. Limited by efficiency limitation, current material weight and adsorption effectiveness, the upper energy density limit of Cha's IA-PEMFC system was around 7 wt.% and 54 g_{H2} L^{-1}. However, these values could be enhanced by efficient system integration, improvement of catalysts and adsorbents, using light tank materials, improving power density of PEMFC, etc.

6.5.4 Direct ammonia fuel cells

There are several types of DAFCs under research, including direct ammonia alkaline fuel cells (DA-AFC) [64–66], direct ammonia molten hydroxide fuel cells (DA-MHFC) [67] and direct ammonia solid oxide fuel cells (DA-SOFC) [1,68–70]. Those DAFCs were all developed from fuel cells with high NH_3 tolerance. Current DAFCs are still at the R&D stage. Among these DAFCs, DA-SOFC operates at higher temperature (400~1000°C) and uses Ni-based catalysts as anode materials, revealing a better compatibility with NH_3 catalytic decomposition. Therefore, DA-SOFC has a faster NH_3 conversion rate than the other two DAFCs operating at relatively low temperature. Currently, 1 kW-class DA-SOFC stack has been successfully demonstrated and continuously operated at ~700 W electric output for ~1000 h with an electric efficiency as high as 57% [70]. The electrochemical performance of DA-SOFC stack is only 3.5% lower than that of SOFC stack fueled with 75% H_2 + 25% N_2 mixture. In contrast, DA-AFC and DA-MHFC are still tested at a lab-level scale in spite of a record power density of 0.45 W cm^{-2} achieved by a DA-AFC in the lab [64]. To promote the utilization of DA-SOFC, several systems have been proposed and evaluated from the perspective of energy assessment [71,72]. Baniasadi and Dincer [71] proposed a combined heat and power (CHP) system integrating a proton-conducting DA-SOFC and micro-gas turbine applicable for vehicles as Fig. 6.13A shows. Energy efficiency and exergy efficiency were both evaluated based on this proposed hybrid system. Results revealed that as heat recovery was integrated and applied current varied, energy efficiency was in the range from 40% to 60%, while exergy efficiency was much higher, between 60% and 90%. The largest exergy losses (39~52%) occurred in DA-SOFC at a maximum power output of 50~62 kW, and the exergy losses of heat recovery and micro-gas turbine accounted for 30~34% of all the exergy losses. As operating temperature of DA-SOFC rises from 600°C to 700°C, entropy generation rate dropped by 25%. After comparing 16 different system configurations, they concluded that the system using DA-SOFC sized for top speed and 100% energy regeneration for traction had the lowest fuel consumption, ~0.3 L per 100 km ride. Another DA-SOFC-based hybrid system was proposed by Perna et al. [72] as Fig. 6.13B shows. This DA-SOFC-based system integrated DA-SOFC, catalytic burner, H_2-N_2 separation membrane, H_2 compressor and other auxiliary equipment to co-generate heat, power (for charging EVs) and 100 kg/day H_2 (capable of refueling 20~30 fuel cell vehicles). Two different design concepts were used and compared based on the minimum DA-SOFC size principle: one is the minimum DA-SOFC size for producing required H_2 and generating the electricity needed by the whole system, the other one is the minimum size for producing required H_2 and generating the electricity needed by both the whole system and extra EV charging. The former design resulted in a higher NH_3-to-H_2 efficiency of 70% and a higher overall efficiency of 81% (an NH_3-to-heat efficiency of 11%, no power generation). The latter design led to an NH_3-to-H_2

efficiency of 42%, an NH_3-to-power efficiency of 14% and a lower overall efficiency of 71% (an NH_3-to-heat efficiency of 15%). Provided electric power used for H_2 compression was ignored, the hybrid system has ~12% higher overall efficiency. High efficiency can be recognized for DAFC. However, current feasible economic pattern and scale-up demonstration were still in demand.

6.6 Summary

Ammonia (NH_3) is a promising and carbon-free energy carrier due to high volumetric energy density, easy liquification, mature production, storage and transportation, and cabon-free feature. In this chapter, we described a complete concept and application scene of ammonia-based energy roadmap, and summarized the attraction of NH_3 energy and the related projects. Used for H storage, NH_3 can signficantly reduce the end-user H costs by 41~83% due to cheap storage and transportation and easy purification. Then the clean and efficiency technologies for NH_3 production and NH_3 utilization (NH_3-to-H_2 and NH_3-to-power) were reviewed, and the typical systems were presented and compared. The technical maturity, existing challenges and potential applications of these alternative NH_3-related technologies were discussed.

References

[1] Y. Luo, Y. Shi, S. Liao, C. Chen, Y. Zhan, C. Au, et al. Coupling ammonia catalytic decomposition and electrochemical oxidation for solid oxide fuel cells: a model based on elementary reaction kinetics, J Power Sources 423 (2019) 125–136.

[2] S. Giddey, S.P.S. Badwal, A. Kulkarni, Review of electrochemical ammonia production technologies and materials, Int J Hydrogen Energ 38 (2013) 14576–14594.

[3] S. Dutta, A review on production, storage of hydrogen and its utilization as an energy resource, J Ind Eng Chem 20 (2014) 1148–1156.

[4] C. Zamfirescu, I. Dincer, Ammonia as a green fuel and hydrogen source for vehicular applications, Fuel Process Technol. 90 (2009) 729–737.

[5] C. Zamfirescu, I. Dincer, Using ammonia as a sustainable fuel, J Power Sources 185 (2008) 459–465.

[6] A. Klerke, C.H. Christensen, J.K. Nørskov, T. Vegge, Ammonia for hydrogen storage: challenges and opportunities, J Mater Chem 18 (2008) 2304–2310.

[7] S. Ma, H. Zhou, Gas storage in porous metal—organic frameworks for clean energy applications, Chem Commun 46 (2010) 44–53.

[8] O. Siddiqui, I. Dincer, A review and comparative assessment of direct ammonia fuel cells, Therm Sci Eng Progr 5 (2018) 568–578.

[9] A. Valera-Medina, H. Xiao, M. Owen-Jones, W.I.F. David, P.J. Bowen, Ammonia for power, Prog Energ Combust 69 (2018) 63–102.

[10] A. Afif, N. Radenahmad, Q. Cheok, S. Shams, J.H. Kim, A.K. Azad, Ammonia-fed fuel cells: a comprehensive review, Renew Sustain Energ Rev 60 (2016) 822–835.

[11] X. Wang, L. Li, T. Zhang, B. Lin, J. Ni, C. Au, et al. Strong metal–support interactions of Co-based catalysts facilitated by dopamine for highly efficient ammonia synthesis:in situ XPS and XAFS spectroscopy coupled with TPD studies, Chem Commun 55 (2019) 474–477.

[12] C. Luo, Research and development status of ammonia-hydrogen fuel cell, Sino-Global Energ 23 (2018) 20–26.

[13] Y. Kojima, Ammonia as a hydrogen carrier, J Jpn Institut Energ 95 (2016) 364–370.

[14] Industry analysis and development trend of ammonia synthesis industry, 2019. Available from: https://www.sohu.com/a/300367018_693849.

[15] New energy: costs, advantages and disadvantages of various hydrogen production technologies, 2019. Available from: https://www.sohu.com/a/292077544_99917855.

[16] PPI. Liquid ammonia vane, 2020. Available from: http://yean.100ppi.com/.

[17] Simon E. Green ammonia. REFUEL Kickoff Meeting, Denver, 2017. Available from: <https://arpa-e.energy.gov/sites/default/files/04c%20Denver-Green%20Ammonia-Siemens-final.pdf.

[18] Brown T. Green ammonia pilot plants now running, in Oxford and Fukushima, 2018. Available from: https://ammoniaindustry.com/green-ammonia-pilot-plants-now-running-in-oxford-and-fukushima/.

[19] JGC Corporation. World's First Successful Ammonia Synthesis Using Renewable Energy-Based Hydrogen and Power Generation, 2018. Available from: https://www.jgc.com/en/news/assets/pdf/20181019e.pdf.

[20] Alkammonia. Ammonia-fuelled alkaline fuel cells for remote power applications. Available from: http://alkammonia.eu/.

[21] Brown T. The AmVeh—an ammonia fuelled car from South Korea NH3 Fuel Association, 2013. Available from: https://nh3fuelassociation.org/2013/06/20/the-amveh-an-ammonia-fueled-car-from-south-korea/.

[22] Science and Technology Promotion Agency Public Relations Division. Establishment of "green ammonia" consortium. JST, 2017. Available from: https://www.jst.go.jp/osir-ase/20170725/index.html.

[23] Advanced Research Project Agency-Energy (ARPA-E). Renewable Energy to Fuels Through Utilization of Energy-Dense Liquids (REFUEL), 2016. Available from: https://arpa-e.energy.gov/sites/default/files/documents/files/REFUEL_Project_Descriptions_Final.pdf.

[24] Undertaking FCAH. Major project to convert offshore vessel to run on ammonia-powered fuel cell, 2020. Available from: https://www.fch.europa.eu/sites/default/files/Press%20release%20ShipFC%20project%20%28004%29.pdf.

[25] Energy R. Ammonia producers dominate hydrogen pilot projects in Australia, 2020. Available from: https://www.rystadenergy.com/newsevents/news/newsletters/sera-archive/renewable-april-2020/.

[26] Brown T. Green ammonia dominates hydrogen demonstrations in Australia, 2020. Available from: https://www.ammoniaenergy.org/articles/green-ammonia-dominates-hydrogen-demonstrations-in-australia/.

[27] Eur Commiss. Best available techniques for the manufacture of large volume inorganic chemicals—Ammonia, Acids and Fertilisers. European Page of Commission Integrated Pollution Prevention and Control; 2007. Technical Report August.

[28] C. Smith, A.K. Hill, L. Torrente-Murciano, Current and future role of Haber–Bosch ammonia in a carbon-free energy landscape, Energ Environ Sci. 13 (2020) 331–344.

[29] S. Pimputkar, S. Nakamura, Decomposition of supercritical ammonia and modeling of supercritical ammonia–nitrogen–hydrogen solutions with applicability toward ammonothermal conditions, J Supercrit Fluid 107 (2016) 17–30.

[30] L. Forni, D. Molinari, I. Rossetti, N. Pernicone, Carbon-supported promoted Ru catalyst for ammonia synthesis, Appl Catalys 185 (1999) 269–275.

[31] D.E. Brown, T. Edmonds, R.W. Joyner, J.J. McCarroll, S.R. Tennison, The genesis and development of the commercial bp doubly promoted catalyst for ammonia synthesis, Catal Lett 144 (2014) 545–552.

[32] H. Liu, Ammonia synthesis catalyst 100 years: practice, enlightenment and challenge, Chin J Catal 35 (2014) 1619–1640.

[33] B. Lin, Y. Guo, R. Liu, X. Wang, J. Ni, J. Lin, et al. Preparation of a highly efficient carbon-supported ruthenium catalyst by carbon monoxide treatment, Ind Eng Chem Res 57 (2018) 2819–2828.

[34] B. Lin, Y. Liu, L. Heng, X. Wang, J. Ni, J. Lin, et al. Morphology effect of ceria on the catalytic performances of Ru/CeO$_2$ catalysts for ammonia synthesis, Ind Eng Chem Res 57 (2018) 9127–9135.

[35] B. Lin, L. Heng, B. Fang, H. Yin, J. Ni, X. Wang, et al. Ammonia synthesis activity of alumina-supported ruthenium catalyst enhanced by alumina phase transformation, ACS Catal 9 (2019) 1635–1644.

[36] B. Lin, Y. Guo, J. Lin, J. Ni, J. Lin, L. Jiang, et al. Deactivation study of carbon-supported ruthenium catalyst with potassium promoter, Appl Catalys 541 (2017) 1–7.

[37] B. Lin, Y. Guo, C. Cao, J. Ni, J. Lin, L. Jiang, Carbon support surface effects in the catalytic performance of Ba-promoted Ru catalyst for ammonia synthesis, Catal Today 316 (2018) 230–236.

[38] B. Lin, K. Wei, J. Ni, J. Lin, KOH activation of thermally modified carbon as a support of Ru catalysts for ammonia synthesis, Chemcatchem 5 (2013) 1941–1947.

[39] Xie K. The national first ruthenium-based low-pressure ammonia synthesis plant in China finishes standardization, 2018. Available from: http://www.stdaily.com/index/kejixinwen/2018-03/08/content_645568.shtml.

[40] The first ruthenium-based low-temperature and low-pressure ammonia synthesis plant in China was successfully in operation by one shot. 2019-07-31. <https://www.sohu.com/a/330696652_120067664>.

[41] Beijing Sanju Environmental Protection and New Materials Co. Ltd. Low-energy and high-efficiency ruthenium-based ammona synthesis technology. Available from: http://www.sanju.cn/Home/nav/116.html.

[42] H. Zhang, L. Wang, J. Van Herle, F. Maréchal, U. Desideri, Techno-economic comparison of green ammonia production processes, Appl Energ 259 (2020) 114135.

[43] T. Grundt, K. Christiansen, Hydrogen by water electrolysis as basis for small scale ammonia production A comparison with hydrocarbon based technologies, Int J Hydrogen Energ 7 (1982) 247–257.

[44] C.J. Pickett, J. Talarmin, Electrosynthesis of ammonia, Nature 317 (1985) 652–653.

[45] R.Z. Sørensen, J.S. Hummelshøj, A. Klerke, J.B. Reves, T. Vegge, J.K. Nørskov, et al. Indirect reversible high-density hydrogen storage in compact metal ammine salts, J Am Chem Soc 130 (2008) 8660–8668.

[46] M. Koike, H. Miyagawa, T. Suzuoki, K. Ogasawara, Ammonia as a hydrogen energy carrier and its application to internal combustion engines Sustain- able vehicle technologies: driving the green agendaWoodhead Publishing Limited, Gaydon, Warwichshire, UK, 2012, pp. 61–70.

[47] Nikkei Asian Review. Japanese utilities team on CO$_2$-reducing tech for coal plants. JST, 2017. Available from: https://asia.nikkei.com/Business/Deals/Japanese-utilities-team-on-CO2-reducing-tech-for-coal-plants.

[48] Duynslaegher C. Experimental and numerical study of ammonia combustion. Katholieke Universiteit Leuven, Leuven, 2011.

[49] C.S. Mørch, A. Bjerre, M.P. Gøttrup, S.C. Sorenson, J. Schramm, Ammonia/hydrogen mixtures in an SI-engine: Engine performance and analysis of a proposed fuel system, Fuel 90 (2011) 854–864.

[50] M.F. Ezzat, I. Dincer, Comparative assessments of two integrated systems with/without fuel cells utilizing liquefied ammonia as a fuel for vehicular applications, Int J Hydrogen Energ 43 (2018) 4597–4608.

[51] M. Keller, M. Koshi, J. Otomo, H. Iwasaki, T. Mitsumori, K. Yamada, Thermodynamic evaluation of an ammonia-fueled combined-cycle gas turbine process operated under fuel-rich conditions, Energy 194 (2020) 116894.

[52] M.F. Ezzat, I. Dincer, Energy and exergy analyses of a novel ammonia combined power plant operating with gas turbine and solid oxide fuel cell systems, Energy 194 (2020) 116750.

[53] J.H. Kim, D.H. Um, O.C. Kwon, Hydrogen production from burning and reforming of ammonia in a microreforming system, Energ Convers Manage 56 (2012) 184–191.

[54] ISO 14687-2. Hydrogen fuel—Product specification. International Organization For Standardization. Switzerland, 2012.

[55] J. Cha, Y.S. Jo, H. Jeong, J. Han, S.W. Nam, K.H. Song, et al. Ammonia as an efficient COX-free hydrogen carrier: fundamentals and feasibility analyses for fuel cell applications, Appl Energ 224 (2018) 194–204.

[56] H.M.A. Hunter, J.W. Makepeace, T.J. Wood, O.S. Mylius, M.G. Kibble, J.B. Nutter, et al. Demonstrating hydrogen production from ammonia using lithium imide—Powering a small proton exchange membrane fuel cell, J Power Sources 329 (2016) 138–147.

[57] L. Chmielarz, M. Jabło ska, Advances in selective catalytic oxidation of ammonia to dinitrogen: a review, Rsc Adv 5 (2015) 43408–43431.

[58] K.E. Lamb, M.D. Dolan, D.F. Kennedy, Ammonia for hydrogen storage: a review of catalytic ammonia decomposition and hydrogen separation and purification, Int J Hydrogen Energ 44 (2019) 3580–3593.

[59] L. Lin, Y. Tian, W. Su, Y. Luo, C. Chen, L. Jiang, Techno-economic analysis comprehensive optimization of an on-site hydrogen refuelling station system using ammonia hybrid hydrogen purification with both high H_2 purity high recovery, Sustain Energ Fuels 4 (2020) 3006–3017.

[60] H. Li, A. Caravella, H.Y. Xu, Recent progress in Pd-based composite membranes, J Mater Chem A 4 (2016) 14069–14094.

[61] J. Liu, X. Ju, C. Tang, L. Liu, H. Li, P. Chen, High performance stainless-steel supported Pd membranes with a finger-like and gap structure and its application in NH_3 decomposition membrane reactor, Chem Eng J 388 (2020) 124245.

[62] B. Cox, K. Treyer, Environmental and economic assessment of a cracked ammonia fuelled alkaline fuel cell for off-grid power applications, J Power Sources 275 (2015) 322–335.

[63] Y.A. Gomez, A. Oyarce, G. Lindbergh, C. Lagergren, Ammonia contamination of a proton exchange membrane fuel cell, J Electrochem Soc 165 (2018) F189–F197.

[64] S. Gottesfeld, The direct ammonia fuel cell and a common pattern of electrocatalytic processes, J Electrochem Soc 165 (2018) J3405–J3412.

[65] S. Suzuki, H. Muroyama, T. Matsui, K. Eguchi, Fundamental studies on direct ammonia fuel cell employing anion exchange membrane, J Power Sources 208 (2012) 257–262.

[66] Y. Zhao, B.P. Setzler, J. Wang, J. Nash, T. Wang, B. Xu, et al. An efficient direct ammonia fuel cell for affordable carbon-neutral transportation, Joule 3 (2019) 2472–2484.

[67] J. Yang, H. Muroyama, T. Matsui, K. Eguchi, Development of a direct ammonia-fueled molten hydroxide fuel cell, J Power Sources 245 (2014) 277–282.

[68] M. Ni, M.K.H. Leung, D.Y.C. Leung, Ammonia-fed solid oxide fuel cells for power generation: a review, Int J Energ Res 33 (2009) 943–959.

[69] Q. Ma, R. Peng, Y. Lin, J. Gao, G. Meng, A high-performance ammonia-fueled solid oxide fuel cell, J Power Sources 161 (2006) 95–98.

[70] M. Kishimoto, H. Muroyama, S. Suzuki, M. Saito, T. Koide, Y. Takahashi, et al. Development of 1 kW-class ammonia-fueled solid oxide fuel cell stack, Fuel Cells 20 (2020) 80–88.

[71] E. Baniasadi, I. Dincer, Energy and exergy analyses of a combined ammonia-fed solid oxide fuel cell system for vehicular applications, Int J Hydrogen Energ 36 (2011) 11128–11136.

[72] A. Perna, M. Minutillo, E. Jannelli, V. Cigolotti, S.W. Nam, J. Han, Design and performance assessment of a combined heat, hydrogen and power (CHHP) system based on ammonia-fueled SOFC, Appl Energ 231 (2018) 1216–1229.

Chapter 7

Power balance and dynamic stability of a distributed hybrid energy system

Yu Luo[a], Yixiang Shi[b] and Ningsheng Cai[b]

[a]*National Engineering Research Center of Chemical Fertilizer Catalyst (NERC-CFC), Fuzhou University, Fujian, China;* [b]*Department of Energy and Power Engineering, Tsinghua University, Beijing, China*

7.1 Introduction

Renewable energy is regarded as one of main primary energy sources in the future Energy Internet because of its cleanliness and sustainability. However, the intermittence and fluctuation may lead to power unbalance and power quality degradation, limiting the utilization of renewable energy. To solve these issues, distributed energy storage systems are in demand to improve the dynamic behavior of the distributed energy systems and balance between power supply and demand, which has been mentioned in previous chapters. In Chapter 2, we briefly introduced the main classification of current energy storage technologies. In Chapter 3, we gave a more specific description of power-to-gas (PtG) or power-to-liquid (PtL) technologies from the perspective of technical characteristics, working principles and classification. In Chapter 5, we discussed some case studies of these PtG or PtL technologies on the energy efficiency and economic feasibility. To evaluate the impact of the time-dependent factors such as renewable power and user loads on the distributed energy systems, we should focus more on dynamic behavior and operation of the distributed energy systems in a smaller time-scale (second- to minute-level). On the one hand, a dynamic simulation platform can offer a powerful tool to understand dynamic system integration and optimize the operating strategy for the distributed hybrid systems (DHSs). On the other hand, some typical indicators are necessary to describe the power balance and dynamic stability for the behavior estimation of the DHSs in dynamic operation.

In this chapter, we will give an introduction about a dynamic simulation platform developed by our group [1,2]. This platform includes a dynamic semiphysical model library, enabling to integrate a number of energy conversion devices and other auxiliary components described by physical model

Hybrid Systems and Multi-energy Networks for the Future Energy Internet.
http://dx.doi.org/10.1016/B978-0-12-819184-2.00007-9

or semiphysical model with the validation of the experimental data. We will give a brief review on the methodologies for these component models. Using this dynamic simulation platform, we propose a DHS with high-capacity wind power integration, and analyze its dynamic behavior and operating strategies. According to the essencial characteristic of power balance between supply and demand, a novel indicator, named mutual information (MI), is used to evaluate our proposed DHS in dynamic operation.

7.2 Dynamic system simulation platform

A complete distributed energy system consists of key energy conversion devices, auxiliary components, and control module to form well-organized energy flows. The methodology for developing the dynamic system simulation platform is summarized in brief as follows [2].

7.2.1 Model library

According to the requirement of the dynamic simulation, we can use different methodologies to build the models with different complexities. The component models can be approximately classified into three types, that is, empirical model, numerical model, and abstract model. Empirical models are the simplest models among these three types of models. Although the empirical model is very convenient and with extremely low computational complexity, their accuracy is relatively low, and less physical insight was contained in them. The numerical models are widely used in different fields in different stages of research and development. The numerical models are commonly described by coupled partial differential equations (PDEs) to describe species conservation, charge conservation, as well as energy conservation. Numerical models are commonly used in evaluation of different structure design and to help understand the impact of the operating conditions on device performance and parameter distributions. By given a wide range of possible design considerations and operating parameters, the design could be examined and optimized. In fact, the numerical model is the most accurate and has great utility in some application such as the design and optimization models of fuel cells and batteries, and also could represent the detailed physico-chemical characteristics inside the electrochemical devices. However, they are the slowest to produce predictions and the hardest to configure, providing limited analytical insight for system designers. In addition, the numerical model always needs a number of parameters and some of which are very difficult to be achieved by experiments or from published literature. Thus, model calibration and parameter estimation processes are usually needed. Instead of modeling the dynamic behavior either by describing the detailed physical processes or by empirical approximation, abstract models attempt to provide an equivalent representation such as the electrical-circuit model and the discrete-time model. These models are particularly useful when compatible models of other system components. Although the number of parameters is not large, such models also employ lookup tables that require considerable effort to configure. In addition, despite acceptable accuracy and computational complexity, these models have limited utility for automated design space exploration because they lack analytical expressions for many variables of interest.

In the system level, we focus more on the behavior of the whole system, and less on a certain component. Consequently, the models in the system simulation generally allow more simplifications in the premise that the model is capable of simulating the key techinical characteristics we concern. For example, we may be more insterested in the distributions of various physical fields like temperature, pressure, gas composition, etc, within a certain energy device in a detailed component model. But it could be not a must in a system-level model. When we concern the system efficiency, we require the model to predict the energy input and output. When we concern the dynamic behavior, we need the model to reveal the dynamic response of an energy device to the time-dependent parameters.

7.2.1.1   Fluctuant renewable energy and user loads

In a real distributed energy system, renewable power (like wind power and solar power) and user loads (electrical power load, heat load, cold load, and gas load) are variable with time, weathers, seasons, locations, etc. In a system-level simulation, some simplification can be made for simulating the renewable power input and user loads. Some studies concentrated more on the system response to the variation of power input or user loads, hence, they simplified these variable input or output to random signals fluctuating within a certain range [2,3]. Some studies put more attention on a certain application scene within a typical period, such as PV power with diurnal rhythm, user loads over 1 day in different seasons, etc. In these cases, typical data obtained from actual devices or systems can be used. Even in some cases, we can combine these two methods, that is, adding a certain fluactuation to a periodic signal. Fig. 7.1 shows typical wind power, PV power and end-user power loads [4–6]. Solar power reveals apparent diural rhythm but slightly varies with days. Wind power has a larger variation in a short-term period. The user loads have slight diural rhythm, but varies a lot in the peak-load period every day. This shows a higher uncertainty in the acitivities of human beings.

7.2.1.2   Heat engines

In physics, the description of heat engines such as internal combustion engines (ICEs) should consider a series of complex processes including combustion, compression/expansion, heat recovery, and power generation. These processes can be described in the ICE module, but increase the model complexity. Another particle simplification is to use an empirical model by summarizing and fitting

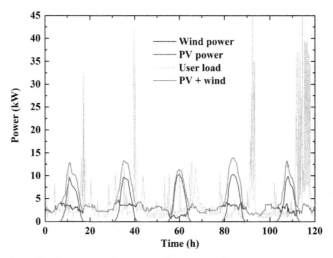

FIGURE 7.1 Typical renewable power output and user loads [4–6].

the experimental data. Ref. [7] summarized a series of experimental data obtained from the ICEs with a wide range of capacities, types, feeding fuels, and revolving speeds, and fitted them using generalized empirical formulations as follows [7]:

$$\eta_0 = 1.3689 \times 10^{-5} P_0(\text{kW}) + 0.3655 \tag{7.1}$$

$$T_{out}^0(\text{K}) = 802.28 - 0.0156 \cdot P_0(\text{kW}) \tag{7.2}$$

$$G_{out}^0\left(\text{m}^3/\text{s}\right) = 3.788 \times 10^{-3} \cdot P_0(\text{kW}) \tag{7.3}$$

$$\eta_{ICE} = \eta_{EG} \cdot \eta_0 \cdot (0.13 + 2.47\tilde{P} - 1.6\tilde{P}^2) \tag{7.4}$$

$$T_{out} = (T_{out}^0 - 273.15)(0.53 + 0.38\tilde{P} + 0.09\tilde{P}^2) + 273.15 \tag{7.5}$$

$$G_{out} = G_{out}^0\left(0.968 + 0.029\tilde{P}\right) \tag{7.6}$$

$$\tilde{P} = P / P_0 \tag{7.7}$$

where P_0 and P are the design power and actual power in kW, respectively. η_0 and η_{ICE} are the electrical efficiencies in the design condition and actual condition, and η_{EG} is the electric generator efficiency. T_{out}^0 and T_{out} are the exhaust gas temperature in Kelvin degree in the design condition and actual condition. G_{out}^0 and G_{out} are the exhaust gas flowrate (m³/s) in the design condition and actual condition. In dynamic operation, the dynamic behavior of the ICE can be described using a first-order lag [3]:

$$G_{ICE}(s) = \frac{1}{\tau_{ICE}s+1} \tag{7.8}$$

Here, τ_{ICE} is the first-order time constant of the ICE. Ref. [3] set the time constant to 2 s.

7.2.1.3 Burners

Typically, we use a lumped continuous stirred reactor (CSTR) model to describe a burner. By feeding excess air, the lumped burner model can calculate the produced gas composition after complete combustion and the corresponding adiabatic temperature through the enthalpy change. Considering the volume and heat capacity of the combustor, we can describe the dynamic behavior of the burners using differential equation of first order with respect to time series.

7.2.1.4 Fuel cells/electrolyzers

Fuel cells and electrolyzers are the reversed processes of each other. They even involve more complex chemical/physical processes, including electrochemical/ chemical reactions, gas diffusion, flow fluid, and heat transfer. Specifically, our group developed a multiscale model methodology to describe fuel cell stacks or electrolyzer stacks from three scales: positive electrode-electrolyte-negative electrode (PEN) scale, single-cell scale, and stack scale [2,8]. Bao et al. [8] has summarized different submodels applicable for these three scale based on a solid oxide fuel cell (SOFC), as Table 7.1 shows. A similar model framework can be used for other types of fuel cells or electrolyzers.

Ref. [2] used this multiscale electrochemical model to describe a reversible SOFC (RSOC) operable in both fuel cell mode and electrolyzer mode. In the PEN scale, we use Arrhenius-form equations to describe electrochemical reaction kinetics and Fick's law to describe gas diffusion. In the single-cell scale, we assume that the single fuel cell is a CSTR, then can simplify a single cell into a lumped cell model. In the stack scale, we link these single cells in parallel or series for scale-up. This multiscale electrochemical model is capable of predicting the electrochemical performance of the RSOC at various operating conditions, even in dynamic operation, as Fig. 7.2 shows.

One simplified method describing a fuel cell or electrolyzer is to use polarization curves to obtain the energy flows within an electrochemical device. As Fig. 7.2A reveals, the voltage for zero current is the open circuit voltage (OCV). RSOC operates in electrolyzer mode (SOEC) when the operating voltage is above OCV, while in fuel cell mode (SOFC) when the operating voltage is lower than OCV. According to the current density, the consumption rate of reactants and the production rate of products can be estimated in the premise of the 100% Faraday efficiency assumption. In the fuel cell mode, fuels are

TABLE 7.1 Different submodels applicable for SOFCs [8].

Scale	Submodel type	Characteristics, assumptions and limitations
PEN scale	Control volume (CV)	• Electrochemical and chemical reactions occur within the whole porous electrode • Isothermal, steady-state and 1D model • Compatible with general or advanced PEN model
	Interface	• Chemical reactions and electrochemical reactions occur at the anode/channel interface and the electrode/electrolyte interface, respectively • Other features are similar to the CV submodel
	Approximate analytical solution	• Approximation of CV submodel with binary reactants • Especially suitable for control-oriented analysis and hardware-in-the-loop simulation
	Advanced PEN model	• Improvement at high fuel utilization by introducing surface chemistry, interface electrochemistry and surface diffusion • Remain the simple frame as general PEN model
Single cell scale	Quasiequilibrium	• Steady-state model • Equilibrium assumption for bulk chemical reactions • The same temperature of outlet gases and solid phase • Compatible with the PEN-scale Interface submodel
	Lumped	• CSTR dynamic model • The same temperature of outlet gases and solid phase • Compatible with all the PEN-scale models
	Distributed	• Distributed dynamic model • Capability of predicting the parameter distributions along gas stream direction • detailed radiant heat transfer with analytical view factors • Compatible with all the PEN-scale models
Stack scale	Identical to cell-level model	• Neglect difference between single cells • Used for the analysis of stack stage or stack network

consumed to generate electricity and heat. The heat value (HV, equal to the enthalpy change) of the consumed fuel denotes the total energy input. Electricity generation can be obtained from the polarization curves (the product of current density and voltage), and the remaining energy is converted to heat.

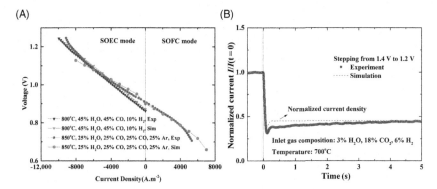

FIGURE 7.2 Multiscale RSOC model validation. *(Modified from Ref. [1]. Copyright 2017 Elsevier Ltd.)*

In the electrolysis mode, the energy flow is more complex with the operating voltage due to the thermal neutral voltage (TNV). When the operating voltage is between OCV and TNV, electricity and heat are consumed to generate fuels. But at higher than TNV, electricity is consumed to generate fuels and heat. The electricity consumption rate can be calculated from the polarization curves, and the difference between the electricity consumption and the HV of the produced fuels determines the required or generated heat. Ref. [9] uses this simplified method to simulate two low temperature electrolyzers, that is, proton exchange membrane electrolysis cell (PEMEC) and alkaline electrolysis cell (AEC). This model uses the polarization curves from the existing literature, and the operating conditions of the selected polarization curves are served as the model input. This model is capable of describing the energy flow, but limited by the variation in operating conditions. Consequently, we can select the models with different complexity to meet the simulation requirement and study target.

7.2.1.5   Batteries

Even though chemical energy storage using PtG or PtL is specifically reviewed in Chapter 5, hybrid energy storage system can significantly enhance the storage characteristics. For example, fuel cells or electrolyzers usually have relatively slow response to electrical power input (second-level). The peak power of the fuel cells or electrolyzers is also limited for longer lifetime. Battery storage is a typical representative of electrical energy storage with fast response and high peak power. Consequently, batteries show high complementary with fuel storage based on PtG or PtL.

Batteries are also a complex electrochemical device. Similarly, a physical battery model should consider charge transfer, mass transfer, heat transfer, as well as electrochemical reactions. A detailed physical model is used in our previously built dynamic system model [2]. This model reveals high accuracy and is capable of predicting the battery characteristics at various temperature

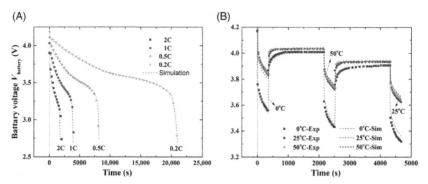

FIGURE 7.3 **Model validation of a Li-ion battery physical model at various charging rate (A) and temperature (B).** *(From Ref. [2]. Copyright 2016 Elsevier Ltd.)*

and charge-discharging current in dynamic operation. In Ref. [2], we calibrated this model and validated it with a $LiMn_2O_4$ battery with a capacity of 11.5 Ah. The comparison between model prediction and experimental results is shown in Fig. 7.3. The battery voltage with time fitted well with experimental data at different charging current and temperature.

Another common battery model is the equivalent circuit model. Compared with the physical model, the equivalent circuit model can significantly simplify the model and reduce the calculation quantity. A well-validated equivalent circuit model can predict the battery dynamic response well, further is applicable for system design, integration, and optimization. The model also should be applicable for multiple battery suppliers and types with easy achieved model parameters. Fig. 7.4 gives a typical equivalent circuit of a battery model. The state-of-charge (SOC) is represented as a function of initial capacity, current history, time, and temperature. Then, the OCV can be calculated by look up table from the SOC value and temperature. The transient response of battery will consider the effects of double layer capacitance and diffusion capacitance (pseudo capacitance). This model takes the advantages of high accuracy, clear physical interpretation and straightforward derivation of model parameters from experimental tests and all of the elements in the circuit are responsible for certain physical-chemical phenomenon. But the self-discharge process was neglected, thus, the prediction for the long-term operation is less accurate.

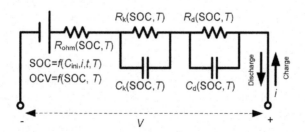

FIGURE 7.4 **A general equivalent circuit model for batteries.**

Similarly to the stack-scale submodel of the fuel cells or electrolyzers, the battery model can be connected in parallel or series to scale up in the premise of the uniform assumption.

7.2.1.6   Catalytic reactors

Catalytic reactors, particularly gas-solid catalytic reactors, are also commonly used in the DHSs when natural gas or other hydrocarbon-based fuels feed to fuel cells or renewable energy sources are stabilized in the hydrocarbon fuels or chemicals. Typical catalytic reactors include steam reforming reactors, methanation reactors, F-T reactors, etc. Several models with different complexity are available for the system simulation, involving the equilibrium model, the lumped CSTR model and the distributed model.

The equilibrium model is the simplest reactor model, where the reactions are assumed to be in the equilibrium state. When we don't concern much about the reactor size, we can assume the reactions to occur fully so that the product gas composition is almost equal to the equilibrium gas composition. The equilibrium model can be isothermal or nonisothermal according to the description of the energy conservation. When we concern about the extent of the catalytic reactions, the reaction kinetics should be put into consideration. The lumped CSTR model is a kinetic model and capable of predicting the dynamic behavior of the reactor. The reaction kinetics are described in the Arrhenius form, and the dynamic behavior of mass transfer and heat transfer can be predicted using the time-dependent differential equations. For the distributed model, the space-dependent and time-dependent terms are both considered in the PDE based on the lumped CSTR model. The distributed model can help to design the reactor from the perspective of operating condition and reactor configuration.

7.2.1.7   Heat exchangers

Different heat exchanger models are available for the system simulation. The most common model is the ε-NTU model, which bridges analytical relationship between the heat exchanger effectiveness (ε) and the number of heat transfer units (NTU). The ε-NTU model is a zero-dimension model. Given the convective heat transfer coefficients, effectiveness and flow mode, the outlet temperature of hot fluid and cold fluid can be determined according to the total heat transfer area. The major flow modes of the heat exchangers include co-flow, counter flow and cross flow. The common tube-shell heat exchangers usually select the counter-flow mode, and some compactly designed heat exchangers like the recuperator of gas turbines or vehicular radiators also use cross-flow mode [10].

In our group's previous study, we also developed a distributed dynamic model for describing the temperature field along with the flow direction in a heat exchanger in coflow or counterflow mode [10]. In this model, the time- and space-dependent PDEs are used to describe the temperature field in both sides (hot fluid side and cold fluid side).

7.2.1.8 Connections, dispatch and control

Some common auxiliary components, such as mixers, splitters, pipelines, valves, and controllers, are necessary to integrate the key components into one complete system. Some common auxiliary components can directly call the intrinsic models of the simulation software.

Assuming uniform mixing, the lumped model can be used for mixers. The time-dependent differential equations can describe the transient gas composition and temperature in the outlet of a mixer. Similarly, the splitters can be also described using a similar lumped model.

Common PID controllers can be used to control the operating conditions of key components to match with transient energy inputs and energy demands, and make the whole system operating more efficiently and economically. The PID control model follows the equation below:

$$u(t) = K_P e(t) + K_I \int_0^t e(\tau) d\tau + K_D \frac{de(t)}{dt} \tag{7.9}$$

where u is the controlled variable and e is the difference between the set value and actual value of the target variable. K_P, K_I, and K_D are three key parameters for PID control, denoting the proportional gain, integration gain and differential gain, respectively. The PID controller is a distributed control with a single input and a single output. Multiobject control can integrate multiple PID controllers, or use other advanced control approaches. For the system-scale control, the control strategies are a significant concern. We can simplify the system-scale cooperative controller to the control signal or command to regulate the energy flow distribution in the whole system.

7.2.1.9 System simulation platform

As mentioned in Chapter 2, the key energy devices for various energy conversion or utilization can be classified into four subsystems: energy generation subsystem, energy storage subsystem, energy recovery subsystem and energy end-use subsystem. The major components included in our model library can be also classified into one of these four subsystems. In our previous study, we built a dynamic simulation platform for a DHS based on the gPROMS software as Fig. 7.5 reveals [2]. This DHS used wind energy and natural gas as the primary energy sources, and integrated a renewable power generator and an ICE as the energy generation subsystem, an RSOC stack and a lithium-ion battery unit as the hybrid energy storage subsystem and end-user power loads as the end-user subsystem. The heat exchangers were integrated in the RSOC subsystem for heat recovery. In Fig. 7.5A, a power manager in the center of the system schematic was used as a cooperative controller to manage the energy flows within the whole system.

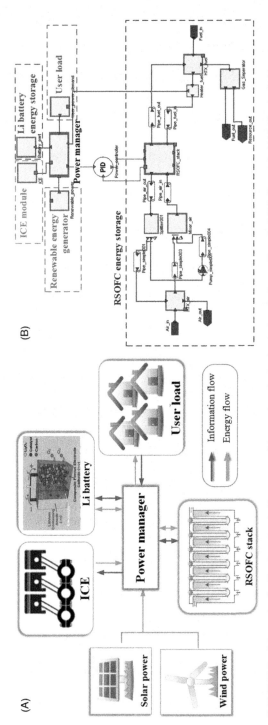

FIGURE 7.5   **Schematic of a distributed hybrid system and its topology in the simulation platform.** *(From Ref. [2]. Copyright 2016 Elsevier Ltd.)*

7.3 Renewable power integration and power balance

7.3.1 Evaluation of the key indicators

To evaluate a DHS, we quantified renewable power integration φ_{RE} and power unbalance ε according to the following equations [2]:

$$\varphi_{RE} = \frac{\int_t P_{direct}^{RE}\, dt + \int_t P_{store}^{RE}\eta_{clc}\, dt}{\int_t P_{supply}\, dt} \tag{7.10}$$

$$\varepsilon = \max_t \left| \frac{P_{demand}(t) - P_{supply}(t)}{P_{demand}(t)} \right| \tag{7.11}$$

As Eq. (7.10) reveals, φ_{RE} has a more generalized implication, denoting the ratio of the supplied power from renewable power to the total power supply. In this DHS, wind power supplies electrical power to the end-users via two pathways, that is, a direct pathway and an indirect one. The direct pathway is that wind power is directly transmitted to the end-users without extra energy conversion, and this part of wind power is represented by P_{direct}^{RE}. The indirect pathway, that is, energy storage pathway, is that wind power is stored first by energy storage devices and then converted back to electrical power for end-users. This part of wind power is represented by P_{store}^{RE}. In the end-user side, the contribution of the energy storage pathway should consider a round-trip efficiency η_{clc}. The value of φ_{RE} is calculated based on the integral electrical power over a certain period. The power unbalance can be represented by the mismatch between power supply and demand, and its value is equal to the maximum relative error of transient power supply and demand.

In addition, the system efficiency and device capacities are also evaluated. In this hybrid system, we concentrated more on the integration of renewable energy and fossil fuels. Consequently, we didn't consider the efficiency of wind generator, but directly use wind power as an input when evaluating the system efficiency. The energy in fuels can be represented by higher heat value (HHV). The system efficiency η_{tot} can be expressed as below:

$$\eta_{tot} = \frac{\int_t P_{supply}\, dt}{\int_t P_{tot}^{RE}\, dt + \int_t q_{in} \text{HHV}\, dt - \int_t P_{store}^{battery} \text{HHV}\, dt} \tag{7.12}$$

where the numerator denotes the total power supply, and the denominator includes wind power input, natural gas input and chemical energy change in the lithium-ion battery pack. The device capacities are determined according to the peak requirement.

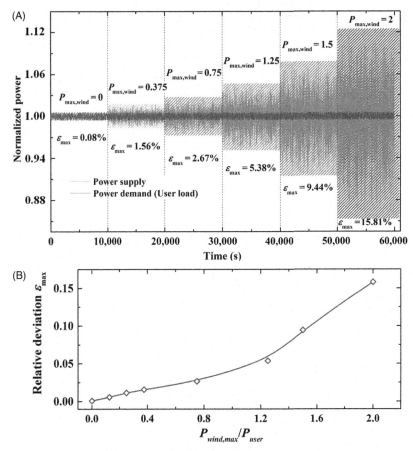

FIGURE 7.6 Power supply and user demand of the system (A), and their relative deviation (B) with different capacity of wind generator when no gas engines are integrated.

7.3.2 Impact of renewable power integration

To study the impact of renewable power integration on a DHS, we use our built simulation platform to develop DHS with difference system schemes. At first, we only integrated wind power generator and reversible solid oxide cells (RSOC) to support electrical power for the end users. Fig. 7.6 shows the power supply and user demand of the system and their relative deviation with wind power capacity varying from 0 to 2. Here, power capacities of each energy devices are normalized by the average power loads. Results indicate that the power supply varies in a large range as wind power capacity rises. Correspondingly, the maximum relative deviation of power supply and demand, representing the degree of power unbalance, significantly increases with wind power penetration rising. In no wind case, the degree of power unbalance is only 0.08%, which means that power supply is quite stable and electricity power quality is very high. When the capacity of wind generator is less than 0.75, the degree of power unbalance almost rises linearly with wind generator capacity. As the capacity of wind generator exceeds 0.75, the degree of power unbalance rises faster. As wind power capacity exceeds user load, the maximum wind power directly sent to user keeps at 1, and the part of wind power that exceeds 1 is stored into gas fuels by RSOC in the SOEC mode. When wind power capacity rises to twice of user load, the degree of power unbalance reaches 15.8%. As a consequence, large wind power penetration greatly increases the degree of power unbalance.

To inhibit the power unbalance caused by large wind power penetration, we integrated an ICE with a capacity of 0.36. Fig. 7.7A shows the corresponding variation in power supply and demand. Here, the maximum wind power penetration was limited below 64% due to the fixed power output of 36% from ICE. When a part of wind power was replaced by stable electrical power, the power unbalance drops from 15.8% by 6.1 points of percentage to 9.7%. Therefore, a larger fraction of stable electrical power in the total power supply can significantly inhibit the degree of power unbalance. Meanwhile, a lithium-ion battery pack was also used to replace RSOC for energy storage in an individual case. The faster response of the lithium-ion battery also remarkably reduced the power unbalance from 9.7% by 7.4 points of percentage to 2.3% at a wind power penetration of 64%. Fig. 7.7B gives the detailed information about the power output of each energy devices. The ICE stably generates an electrical power of 0.36, and the wind power output varies from 0 to the maximum wind power. In the maximum wind power penetration case, the wind power output varies from 0 to 1.24 (1.24 folds of the average power load). When the wind power penetration is lower than 32% (the maximum wind power = 0.64), RSOC or lithium-ion battery pack is required to operate in the SOFC mode all the time. As the wind power penetration further rises, the peak wind power output exceeds to the power loads, hence, RSOC or battery pack also needs to store the surplus wind power sometimes. The power outputs/inputs of RSOC and Li-ion battery reveal that Li-ion battery has a better load follow characteristics than RSOC. However, unlike RSOC, Li-ion battery is not a continuous power generator, even not suitable for long-term energy storage. Long-term power generation or power storage will increase the required capacity significantly, which is not a cost-efficient solution for renewable power integration. Therefore, the DHS needs a specific dynamic operation strategy for better dynamic behavior and higher power balance.

7.3.3 Dynamic operation strategies

Based on the RSOC-based PtG storage, we proposed a series of dynamic operation strategies and evaluated their effects on the DHS. The strategies are described as below:

Strategy 1: Wind power curtailment case. In this case, wind generator is only allowed to supply at most 40% of peak wind power to the end-users.

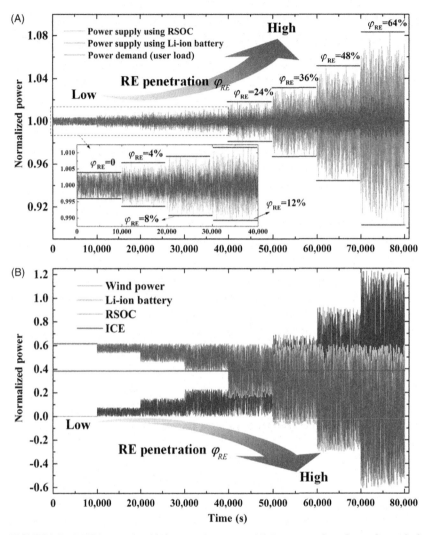

FIGURE 7.7 **DHS integrating 38% gas engine power: (A) Power supply and user demand of the DHS using RSOC and Li-ion battery. (B) Power distribution in these two DHS.** *(From Ref. [1]. Copyright 2017 Elsevier Ltd.)*

RSOCs always operate in fuel cell mode to supply the remaining loads, and no energy storage is used in this case.

Strategy 2: All wind PtG case. In this case, all wind power is converted into gaseous fuel by RSOCs operating in electrolysis mode, and all power loads are supplied by RSOCs operating in fuel cell mode.

Strategy 3: Maximum wind power integration case. In this case, wind power has priority in supply to the end-user power load. The difference between wind power and end-user load is shifted or filled up by RSOCs.

Strategy 4: Partial supply and partial storage case. In this case, wind power has priority in supply at most a limitation ratio γ of the average power load (i.e., $\gamma Pave\ load$) to the end-user power load. The remaining power load is filled up by RSOCs in fuel cell mode. When wind power output exceeds $\gamma PMax\ wind$, a part of RSOCs operate in electrolysis mode to store the excess wind power. In particular, the case is Strategy 2 when $\gamma = 0$, and the case is Strategy 3 when $\gamma = 50\%$.

Strategy 5: Combined storage case. This case is similar to Strategy 4, but the difference is the approach to store wind power. In this case, the excess wind power is stored by the combination of RSOCs and Li-ion batteries. RSOCs in fuel cell mode have priority in supplying at most 48% of the average power load to the end-user in the premise of $SOC_{Li} < 0.85$, and the remaining load is filled up by the Li-ion batteries. When SOC_{Li} is lower than 0.50, the Li-ion batteries have priority in storing the excess wind power. Otherwise, the RSOCs in electrolysis mode are used for wind power storage.

The comparison between various strategies are shown in Fig. 7.8 [2]. When using Strategy 3, the DHS requires RSOCs to shift or fill up the power difference between wind power output and power loads. Therefore, RSOCs operate in a much more fluctuate operating condition. In the maximum wind power integration case, the renewable power integration φ_{RE} and system efficiency η_{tot} are the highest of all the strategies, reaching 56.4% and 54.8%, respectively. However, such a high wind power integration also significantly aggravates the power unbalance (reaching 8.34%). The other strategies offer different options to inhibit the power unbalance. Strategy 1, that is, the wind power curtailment case, is a common approach to enhance power balance. Strategy 1 limits the utilization of wind power, thus, significantly reduces the wind power integration. According to the definition of system efficiency, the waste of wind power leads to the drop in the system efficiency in Strategy 1. However, the power balance only slightly improves when the wind power limitation is 40% of the peak wind power in the wind power curtailment case. Strategy 2, that is, the all wind PtG case, converts all wind power into stable chemical energy, hence, remarkably inhibits the power unbalance below 0.57%. Meanwhile, the wind power integration φ_{RE} and system efficiency η_{tot} are both higher than those in Strategy 1 in spite of twice RSOC capacity requirement. Effective energy storage can dramatically improve the dynamic stability and power balance in a DHS without any wind power curtailment. Nevertheless, stabilizing all wind power through PtG and gas-to-power inevitably cause quite a few power losses due to round-trip efficiency. Therefore, Strategy 4 offers a more flexible strategy to make a tradeoff between power balance and power losses. In Strategy 4, the DHS directly sends a part of wind power to the end-users, and stores the others using RSOCs in electrolysis mode. The remaining power loads are filled up by ICEs and RSOCs in fuel cell mode. Fig. 7.8 gives six cases using Strategy 4, in which the wind power limitations for direct supply to end-users are 0, 10%,

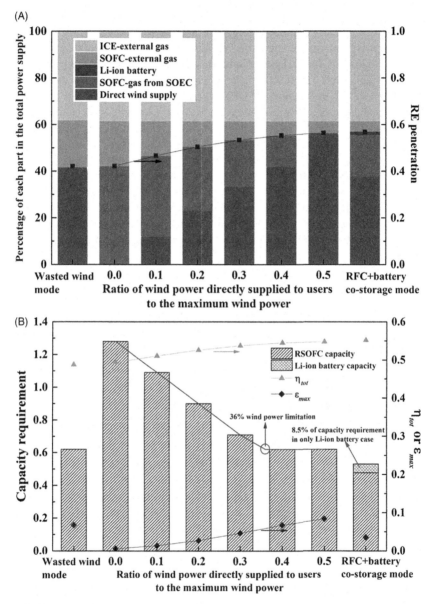

FIGURE 7.8 Comparison between various strategies: (A) Energy consumption distribution and wind power penetration. (B) Power unbalance, system efficiency and required capacity. *(From Ref. [2]. Copyright 2016 Elsevier Ltd.)*

20%, 30%, 40%, and 50% of the peak wind power output, respectively. The first case denotes Strategy 2, while the last one denotes Strategy 3. As the wind power limitation rises, the stored gaseous fuels from RSOCs operating in electrolysis mode decrease, and power unbalance increases while the overall system efficiency enhances. The RSOC capacity requirement in Fig. 7.8B is defined as the total rated power when all RSOCs operate in fuel cell mode. The required RSOC capacity drops first and keeps unchanged with wind power limitation increasing. The critical wind power limitation is 36% of the peak wind power output, and the minimum RSOC capacity requirement is 0.62, that is, 62% of the average power loads. To compromise these factor comprehensively, the DHS may prefer Strategy 4 with an optimal wind power limitation ratio of 36%. In this case, wind power integration can reach as high as 54.6%, the overall system efficiency is 54.2% and power unbalance is 5.95%, meanwhile, the DHS only requires the least RSOC capacity. These factors are much better than those in Strategy 1 (wind power curtailment case). However, the power unbalance is still quite high.

It is hard to optimize all these factors in an DHS by using a single energy storage device. RSOCs with enough storage tanks can seasonally store fuels without zero self-charging losses, but have a peak power output in fuel cell mode (limited power density), and rarely operate at high polarization voltage to ensure enough lifetime. Li-ion battery with a much smaller capacity can suffer transient high power input or output, but self-charging and limited energy density are the drawbacks of the Li-ion batteries. To further reduce power unbalance and improve system dynamic behaviors, combining RSOCs with fast-response Li-ion battery is a quite feasible and complementary energy storage solution for an DHS with high wind power integration. In the meantime, the total requirement for energy storage capacity can be remarkably reduced. Strategy 5 represents the combined energy storage solution in the DHS. Based on Strategy 4 with the optimal wind power limitation ratio, Strategy 5 uses RSOC-battery costorage to replace the single storage of RSOC. The combination of RSOC and Li-ion battery enhances the wind power integration φ_{RE} by 2.3 points of percentage to 56.9% and system efficiency η_{tot} by 1 points of percentage to 55.2%. Meanwhile, the power unbalance also drops from 5.95% by 2.39 points of percentage to 3.56%. This reveals that the power quality is much more acceptable to the end-users, and the system operation dynamic stability is dramatically enhanced. For the energy storage capacity, Strategy 5 only requires 77.5% of the required RSOC capacity in the RSOC-only case and 8.5% of the required Li-ion battery capacity in the battery-only case. RSOCs are used to store most of surplus wind power, and Li-ion batteries with high charge-discharge current are in charge of filling up the insufficient power demand or shifting transient high wind power output. Here, we gave a more intuitive example for energy storage capacity. When the average user loads are 100 kW, the DHS using the optimal RSOC-only storage requires 62 kW RSOC, the one using Li-ion battery-only storage requires 15.7 kWh Li-ion battery, and the one using RSOC + Li-ion

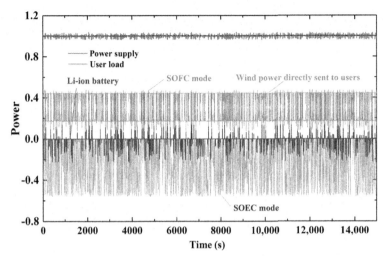

FIGURE 7.9 Transient power of various energy devices in the combined energy storage cases (**Strategy 5**). *(From Ref. [2]. Copyright 2017 Elsevier Ltd.)*

battery hybrid storage only requires 48 kW RSOC and 1.3 kWh Li-ion battery to meet seasonal and fast-response energy storage and power emergent supply.

Fig. 7.9 shows the transient power distributions of various energy devices in the combined energy storage cases. The combined energy storage solution can significantly save the total size of the energy storage devices, and utilize wind power in a better way.

7.3.4 Co-generation of electricity, heat and gas

Dynamic exergy analysis (Second Law of Thermodynamics) was performed for the DHS using RSOC-based PtG storage. Fig. 7.10A shows the variation of

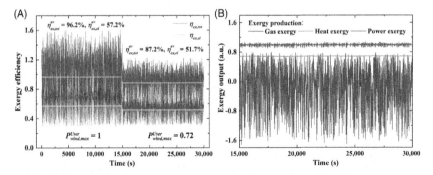

FIGURE 7.10 (A) Variation of transient electrical exergy efficiency and overall exergy efficiency before and after using the optimal energy storage strategy. (B) Transient electricity, heat and gas exergy output after using the optimal energy storage strategy.

transient electrical exergy efficiency and overall exergy efficiency before and after using the optimal energy storage strategy. This electricity load is given, hence this system determines heat output based on electricity. The electrical exergy efficiency is used to evaluate the efficiency of electricity generation, while the overall exergy efficiency is used to evaluate the efficiency of heat/electricity cogeneration. For each 15,000 s, the time-mean electrical exergy efficiency $\eta_{av\ ex,el}$ and overall exergy efficiency $\eta_{av\ ex,tot}$ are calculated based on the integral value of inlet/outlet exergy, and shown in Fig. 7.10A. Fig. 7.10A indicates that $\eta_{ex,el}$ or $\eta_{ex,tot}$ fluctuates more strongly before optimization, but the lower bounds for the values before and after optimization are similar. This is because the lower bound corresponds to the exergy efficiency when the weather is windless. Therefore, the time-mean value is higher than that after optimization. After using the optimal energy storage strategy, the time-mean $\eta_{av\ ex,el}$ drops from 57.2% to 51.7% and $\eta_{av\ ex,tot}$ drops from 96.2% to 87.2%. The optimal energy storage strategy sacrifices electrical efficiency by 4.5% and overall efficiency by 9%, but the system becomes much less fluctuate. Fig. 7.10B shows the transient electricity, heat and gas exergy output during 15,000–30,000 s. The electricity exergy output is controlled by PID controller and varies around the electricity load (0.94–1.06), thus, the time-mean electricity exergy output is equal to 1. The heat exergy output varies less strongly than the other two energy sources because of slower heat transfer process. The time-mean heat exergy is 0.687. The fluctuation of gas exergy output is the strongest of these three energy sources. Utilization and storage of gas fuels depend much on transient wind power. A part of RSOCs always operate in fuel cell mode to consume gas fuels and supply at least 0.28 electricity because of an upper limit of 0.72 for direct wind power supply. As transient wind power is higher than 0.72, a part of RSOCs operate in electrolysis mode to store the part of wind power exceeding 0.72 and enhance the amount of the stored gas fuels. However, the electrical energy storage in the form of gas fuels using RSOCs has to overcome a round-trip efficiency of ~70%. This results only ~0.7 unit of stable electricity can be obtained when 1 unit of wind power is stored and converted back to electricity by the RSOC. As a result, the time-mean gas exergy output is −0.155, which means the system demands an extra natural gas input.

To simultaneously meet user demand for electricity, heat and gas, the output regulation of these three energy sources should be figured out. Once electricity load is fixed, the minimum capacity requirement for power is determined to ensure electricity supply even in windless weather. Fig. 7.11A shows exergy output of electricity, heat and gas at different $P_{wind,max}/P_{load}$. The unit is normalized by the electricity load, hence the exergy output of electricity remains at 1. The exergy output of heat slightly decreases and then recovers with $P_{wind,max}/P_{load}$. The lowest exergy output of heat is found to be at $P_{wind,max}/P_{load} = 2.8$. As $P_{wind,max}/P_{load}$ rises, the wind power that is stored by RSOC operating in the SOEC mode dramatically rises, leading to the time-mean voltage in the SOEC mode rising. The rise in voltage results that SOEC turns from endothermic mode

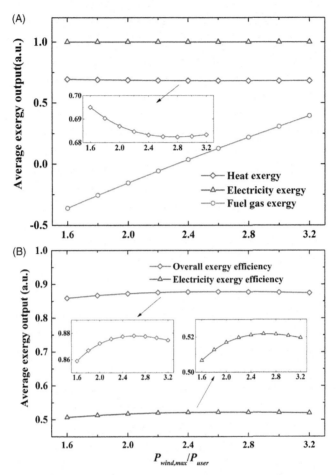

FIGURE 7.11 Exergy output of electricity, heat and gas (A) and exergy efficiency (B) at different $P_{wind,max}/P_{load}$.

to exothermic mode. The exergy output of gas is found to have a nearly linear relation with $P_{wind,max}/P_{load}$. The system consumes gas fuel at $P_{wind,max}/P_{load} < 2.3$, and generates gas fuel at $P_{wind,max}/P_{load} > 2.3$. The exergy output of gas reaches ~ 0.4 at $P_{wind,max}/P_{load} = 3.2$. Gas fuels or electricity can be efficiently converted into heat by using a burner or electric heater, hence, the exergy of heat output can be flexibly regulated according to the heat load. When heat demand is larger than heat output of this system, the part of wind power exceeding 0.72 is prior used for an electric heater to meet heat demand. If wind power is not enough, exact heat can be filled up by a gas burner. When gas demand is large, a large capacity of wind generator is suggested.

Fig. 7.11B shows the variation of the time-mean electrical exergy efficiency and overall exergy efficiency with $P_{wind,max}/P_{load}$. Results indicate and only vary with 2%, and both rise at first then drop slightly. Peak efficiency for and are found to be 52.18% and 87.80% at $P_{wind,max}/P_{load} = 2.6$. Compared with upper limit of direct wind power supply, $P_{wind,max}/P_{load}$ has a much weak dependence on efficiency. Therefore, the regulation of electricity, heat and gas among each other has a little effect on system efficiency. The heat exergy in exhausted gas slightly decreases at first and then recovers at $P_{wind,max}/P_{load} > 2.6$. This study shows a flexible regulation among electricity, heat and gas can be realized by using RSOC in a distributed multienergy system.

7.4 Novel criterion for distributed hybrid systems

In an DHS, power balance requires a match between power supply and demand. The definition of power unbalance ε only reveals the maximum power unbalance in a period, which is still limited in the dynamic operation, especially for the comparison between varied time periods or the evaluation of the cases with transient strong power variation. Our group proposed a novel criterion from the perspective of the information match between power supply and demand, that is, MI [1]. MI is essentially a kind of information enthalpy involving information theory and probability theory, revealing the mutual part of the information entropies two variables contain. Consequently, MI can essentially reflect the general dependence of two variables, which is different from other criteria such as maximum relative deviation or correlation [11–15]. We used MI to quantify the matching degree of power supply and demand. The value of MI $I(S,D)$ can be expressed as below [15–18]:

$$I(S,D) = \sum_D \sum_S P(S,D) \ln \frac{P(S,D)}{P(S)P(D)} \tag{7.13}$$

where $P(S)$ and $P(D)$ denote the probabilities of power supply S and power demand D, respectively, and $P(S,D)$ is the joint probability of power supply and power demand. To quantize the probabilities or joint probability, we discretized the dynamic power supply and demand to two power series according to the time series $\hat{t} = [\Delta t, 2\Delta t, ..., n\Delta t, ...]$:

$$S = \left[P_{supply}(\Delta t), P_{supply}(2\Delta t), ..., P_{supply}(n\Delta t), ... \right] \tag{7.14}$$

$$D = \left[P_{demand}(\Delta t), P_{demand}(2\Delta t), ..., P_{demand}(n\Delta t), ... \right] \tag{7.15}$$

The power supply series and power demand series can be combined into a power supply-demand pair series as below:

$$(S,D) = \left[\begin{matrix} \left(P_{supply}(\Delta t), P_{demand}(\Delta t) \right), \left(P_{supply}(2\Delta t), P_{demand}(2\Delta t) \right), \\ \left(P_{supply}(n\Delta t), P_{demand}(n\Delta t) \right), ... \end{matrix} \right] \tag{7.16}$$

In the following parts, we will give some examples to demonstrate the applications of MI in the distributed energy systems.

7.4.1   Application to evaluate the impact of renewable power integration

In Section 3, the power unbalance has been calculated and used to represent the mismatch degree between power supply and demand. Particularly in Section 3.2, the power unbalance was used to evaluate the impact of renewable power integration. Here, we give an example to use MI for evaluating the impact of renewable power integration [1]. First, we generated power supply-demand pair series as Eq. (7.16) shows and drew them in one rectangular coordinate using power supply as X axis and power demand as Y axis. Fig. 7.12A and B shows the distributions of the power supply-demand pair series at various wind power capacities and energy storage devices, respectively. In both figures, the orange line denotes the exact match between power supply and demand, and each dot denotes one power supply-demand pair for one certain time point. The dot series distribute closer to the orange line, meaning that power supply matches better with power demand. Fig. 7.12A reveals that lower wind power penetration leads to a better match between power supply and demand. When wind power penetration = 64%, the power supply-demand pair dots distribute much more dispersedly, meaning a much higher power unbalance. Fig. 7.12B shows that power supply-demand pair dots obtained from an DHS using RSOC storage distribute more dispersedly than those from an DHS using Li-ion battery storage. Correspondingly, we calculated the values of MI $I(S,D)$ by statistical approach on the basis of these dots. Fig. 7.12C plots the wind power penetration-$I(S,D)$ curves using RSOC storage and Li-ion battery storage, respectively. The match degree, represented by $I(S,D)$, has an inverse relationship with the dispersity of power supply-demand pairs. When wind power penetration = 0, the value of $I(S,D)$ is 2.23 for the RSOC-only storage case and 2.46 for the Li-ion battery-only storage case. As wind power penetration rises, the increasing dispersity of power supply-demand pairs leads to an increasing power unbalance, correspondingly, the value of $I(S,D)$ decreases. At low wind power penetration, $I(S,D)$ is quite sensitive to wind power penetration. When wind power penetration > 24%, $I(S,D)$ decreases much slower with wind power penetration. $I(S,D)$ in the RSOC-only storage case is lower than 0.30, and $I(S,D)$ in the Li-ion battery-only storage case is lower than 0.75 at wind power penetration > 24%. As wind power penetration rises to 64%, $I(S,D)$ is still higher than 0.2 for the RSOC-only storage case, and higher than 0.56 for the Li-ion battery-only storage case. Higher value of $I(S,D)$ in the Li-ion battery case reveals a faster response of Li-ion battery than that of RSOC. This case study reveals that the MI can identify power balance between supply and demand, therefore, be applicable for evaluating wind power penetration and identifying the response features of energy conversion devices.

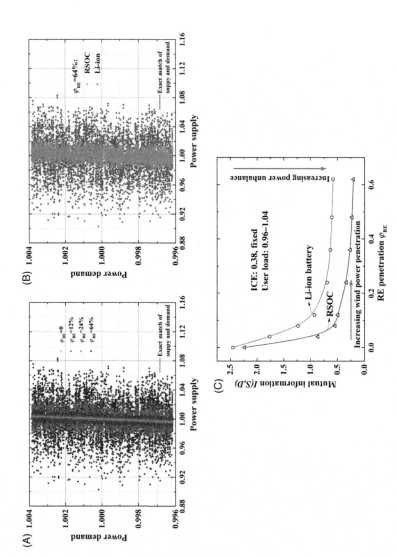

FIGURE 7.12 Distribution of supply-demand pairs for each time-dots: (A) RSOC storage at different wind power penetration. (B) RSOC storage and Li-ion battery storage at wind power penetration = 60%. (C) MI value $I(S,D)$ of the DHS using RSOC and Li-ion battery storage at different wind power penetration. (*From Ref. [1]. Copyright 2017 Elsevier Ltd.*)

7.4.2   Application to detect energy storage capacity

Another application of MI in the DHS is the detection of energy storage capacity. The value of MI varies with energy storage capacity, hence, can be an indicator for detecting energy storage capacity [1]. In the case study, we only used Li-ion battery for energy storage, and its optimal capacity was determined according to the maximum power unbalance mitigation and the minimum capacity principles. For safe and long-lifetime operation, we allowed the Li-ion battery to discharge at SOC > 0.3 and charge at SOC < 0.9. Fig. 7.13A reveals the variation of SOC with time at various Li-ion battery capacities. Here, Li-ion battery capacity was divided by the average user load for normalization, hence, the unit of Li-ion battery capacity is "h". When Li-ion battery capacity = 0.336 h, the SOC of Li-ion battery is always between 0.3 and 0.9, meaning Li-ion battery can always offer enough electrical power and store all the surplus wind power. Fig. 7.13B reveals that power supply match well with power demand (normalized to ~1) when Li-ion battery capacity > 0.157 h. As Li-ion battery capacity reduces to 0.152 h, the peak SOC slightly exceeds the upper limit, that is, 0.9, at around 59,000 s. Correspondingly, Fig. 7.13B appears a peak power supply exceeding 1.2 at the corresponding time when Li-ion battery capacity = 0.152 h, meaning Li-ion battery is unable to store more wind power at this time. As Li-ion battery capacity drops to 0.084 h, Li-ion battery failed to supply enough electrical power or store surplus wind power sometimes. When Li-ion battery capacity = 0.028 h, serious power unbalance frequently occurred, revealing an insufficient energy storage. By observing SOC at all time, we can determine the optimal Li-ion battery capacity to be 0.157 h. However, this is hard to predict a required capacity in advance. We need an indicator to quantify the match between power supply and demand, and further to detect the required capacity in the design stage or the real time operation.

Fig. 7.13C shows the variation of MI $I(S,D)$ and maximum relative deviation of power supply and demand ε_{max} with Li-ion battery capacity. When Li-ion battery capacity > 0.157 h, the value of $I(S,D)$ remains at 0.56, and ε_{max} at 2.3%. When Li-ion battery capacity drops from 0.157 to 0.152 h, $I(S,D)$ rapidly drops from 0.56 to 0.002, and ε_{max} rises from 2% to 50%. As Li-ion battery capacity further drops, ε_{max} rises to 64% at Li-ion battery capacity > 0.141 h and remains at 64% at Li-ion battery capacity < 0.141 h. When Li-ion battery capacity is insufficient, ε_{max} becomes quite insensitive, which is hard to reflect how much energy storage capacity is lacking. As for $I(S,D)$, $I(S,D)$ slightly recovers from 0.002 to 0.08 with Li-ion battery capacity dropping from 0.152 h tp 0.028 h. Consequently, MI is not only quite sensitive to the optimal (critical) Li-ion battery capacity, but also can reflect each stages of Li-ion battery capacity. Apart from energy storage capacity, MI is also applicable to detect the capacity of power generators and other energy conversion devices. MI also has the application potential in the real time system control and diagnosis [1].

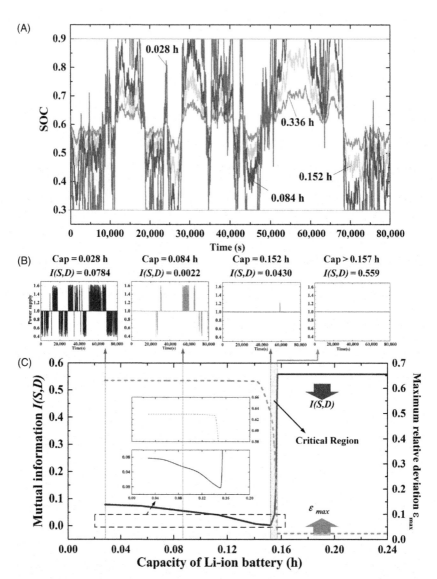

FIGURE 7.13 (A) State-of-charge of Li-ion batteries with various capacities. (B) Values of $I(S,D)$ and ε_{max} at various Li-ion battery capacity and typical dynamic power supply. *(From Ref. [1]. Copyright 2017 Elsevier Ltd.)*

7.4.3 Application to evaluate energy storage strategies

MI can be also used to evaluate energy storage strategies. We used MI to reevaluate various energy storage strategies in Section 3.3. Here, we give a simplified case study. Three energy storage strategies, including RSOC-only storage, Li-ion battery-only storage and RSOC + Li-ion battery hybrid storage,

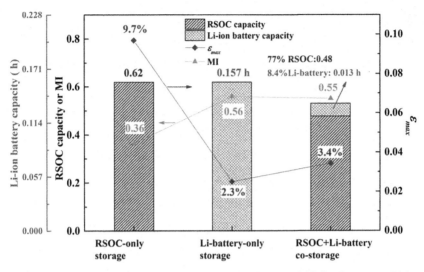

FIGURE 7.14 Values of MI, ε_{max} and storage capacity using RSOC-only storage, Li-ion battery-only storage and RSOC + Li-ion battery hybrid storage. *(From Ref. [1]. Copyright 2017 Elsevier Ltd.)*

were compared in Fig. 7.14 [1]. In Section 3.3, the maximum relative deviation of power supply and demand ε_{max} was used to evaluate power unbalance. Using RSOC-only storage, the DHS has an ε_{max} of 9.7%. The ε_{max} of the DHS using Li-ion battery-only storage is only 2.3%. The DHS using RSOC + Li-ion battery hybrid storage has an ε_{max} of 3.4%. From the perspective of power match between supply and demand, MI analysis revealed that $I(S,D)$ was 0.36 for the DHS using RSOC-only storage, 0.56 for the one using Li-ion battery storage, and 0.55 for the one using RSOC + Li-ion battery hybrid storage. Consequently, MI is capable of representing the match degree between power supply and demand, hence, applicable for evaluating power balance.

7.5 Summary

In this chapter, we mainly focused on dynamic behavior and operation of the distributed energy systems in a shorter timescale. First, we gave an introduction about a self-developed dynamic simulation platform. This platform includes a dynamic semiphysical model library, enabling to integrate a number of energy conversion devices and other auxiliary components with each other into larger energy networks. We gave a brief review on the methodologies for the typical energy conversion component models. Using this dynamic simulation platform, we propose a DHS with high-capacity wind power integration, and analyze its dynamic behavior and operating strategies. Finally, we introduced a novel indicator, that is, MI, and showed three case studies to reveal its applicability in DHSs from the perspective of evaluation of power supply-demand match and detection of energy device capacities.

References

[1] Y. Luo, Y. Shi, Y. Zheng, Z. Gang, N. Cai, Mutual information for evaluating renewable power penetration impacts in a distributed generation system, Energy 141 (2017) 290–303.

[2] Y. Luo, Y. Shi, Y. Zheng, N. Cai, Reversible solid oxide fuel cell for natural gas/renewable hybrid power generation systems, J Power Sources 340 (2017) 60–70.

[3] T. Senjyu, T. Nakaji, K. Uezato, T. Funabashi, A hybrid power system using alternative energy facilities in isolated island, IEEE Trans Energy Convers 20 (2005) 406–414.

[4] C. Chen, S. Duan, T. Cai, Q. Dai, Shor-term photovoltaic generation forecasting system based on fuzzy recognition, Trans China Electrotech Soc 7 (2011) 83–89.

[5] J.D. Maclay, J. Brouwer, G.S. Samuelsen, Dynamic modeling of hybrid energy storage systems coupled to photovoltaic generation in residential applications, J Power Sources 163 (2007) 916–925.

[6] Shi J. Optimization for very short-term wind power forecasting algorithm: North China Electric Power University; 2012.

[7] He X. Typical off-design performances of internal combustion engine CHP system: Institute of engineering thermal physics, Chinese Academy of Sciences; 2008.

[8] C. Bao, Y. Shi, E. Croiset, C. Li, N. Cai, A multi-level simulation platform of natural gas internal reforming solid oxide fuel cell-gas turbine hybrid generation system. Part I. Solid oxide fuel cell model library, J Power Sources 195 (2010) 4871–4892.

[9] Y. Luo, X. Wu, Y. Shi, A.F. Ghoniem, N. Cai, Exergy analysis of an integrated solid oxide electrolysis cell-methanation reactor for renewable energy storage, Appl Energy 215 (2018) 371–383.

[10] C. Bao, N. Cai, E. Croiset, A multi-level simulation platform of natural gas internal reforming solid oxide fuel cell-gas turbine hybrid generation system. Part II. Balancing units model library and system simulation, J Power Sources 196 (2011) 8424–8434.

[11] A.M. Fraser, H.L. Swinney, Independent coordinates for strange attractors from mutual information, Phys Rev A 33 (1986) 1134–1140.

[12] F. Maes, A. Collignon, D. Vandermeulen, G. Marchal, P. Suetens, Multimodality image registration by maximization of mutual information, IEEE Trans Med Imaging 16 (1997) 187–198.

[13] P. Viola, W.M. Wells III, Alignment by maximization of mutual information, Int J Comput Vision 24 (1997) 137–154.

[14] H. Peng, F. Long, C. Ding, Feature selection based on mutual information: criteria of max-dependency, max-relevance, and min-redundancy, IEEE Trans Pattern Anal 27 (2005) 1226–1238.

[15] R. Steuer, J. Kurths, C.O. Daub, J. Weise, J. Selbig, The mutual information: detecting and evaluating dependencies between variables, Bioinformatics 18 (2002) S231–S240.

[16] C.E. Shannon, The mathematical theory of communication, Bell Syst Tech J 27 (1948) 379–424 623–656.

[17] A. Kolmogorov, Logical basis for information theory and probability theory, IEEE Trans Inform Theory 14 (1968) 662–664.

[18] T.M. Cover, J.A. Thomas, Elements of information theory, Wiley, New York, (1991).

Chapter 8

Applying information technologies in a hybrid multi-energy system

Yu Luo[a], Yixiang Shi[b] and Ningsheng Cai[b]

[a]National Engineering Research Center of Chemical Fertilizer Catalyst (NERC-CFC), Fuzhou University, Fujian, China; [b]Department of Energy and Power Engineering, Tsinghua University, Beijing, China

8.1 Why information technologies are needed?

Hybrid multi-energy system (HMES) is a complex and highly nonlinear system, which is totally different from traditional energy system. In a hybrid energy system, electricity, heat, and gas are all integrated. There exist varies of energy conversions among different forms. Except for the energy flow, information flow is also considered in this conception. The combination of energy and information makes this system even more complicated. Besides the fact that the numbers and forms for system components are various, the function and behavior mode of the participants in hybrid energy system also tends to be more difficult. Traditionally, the consumers in energy system, namely users, only consume energy, while the producers only generate energy. In a novel HMES, however, users have the ability to supply extra energy to market and the energy flow becomes bi-directional. Meanwhile, the increasing use of renewable energy, which has a huge fluctuation and a strong intermittence, results in randomness of HMES.

Due to the complexity in space and the randomness in time, there exist a large number of issues. Since lots of wind energy and solar energy are integrated, the whole system becomes more unsteady, which implies that new control strategies, for example, prediction-based control and distributed methods, should be utilized. In comparison with general energy systems, HMES has significantly larger number of participants from power system, gas network, and heat network. This means that numerous data with various types will be generated. The collection, analyzation, and utilization of these data tend to be another problem. Discovering valuable information behind them is also worthwhile. The next problem lies in energy transaction, with which energy flows effectively

Hybrid Systems and Multi-energy Networks for the Future Energy Internet.
http://dx.doi.org/10.1016/B978-0-12-819184-2.00008-0

in the whole system. Because each participant in HMSE could be either seller or consumer, the transaction management becomes a challenge.

Modern information technologies, including block chain, big data, etc. are regarded as an ideal solution to solve problems of complex system and cope with issues in HMSE. They are capable of collecting data from every component in the system, handling with millions of data and revealing the underlying value, building novel management system. They have the potential to be used in every aspect of energy system, from energy generation to consumption. It is believed that with the application of information technologies, modern energy system will be smarter and more efficient.

8.2 Block chain and energy transaction

Energy, in the form of electricity, heat, and gas, is a kind of commodity widely used in the industry and household. Conventionally, energy is generated by suppliers like power corporations that are generally big enterprises and then transmitted through power grid or heat network. Finally, energy reaches to the user side and is consumed. This is a typical situation when energy transaction happens in the past. Every participant has a fixed role in the whole energy market and electricity, heat, and gas network are separated systems, which are conducted by different companies. The structure of this energy market is rather easy. Energy flows from corporations to users and money flows in an opposite direction.

The occurrence of energy Internet, however, has broken the traditional way for energy transaction. In an energy Internet, multiple forms of energy integrate with each other and conversions happen frequently between different energy forms. A corporation cannot only supply electricity but also heat and natural gas. Energy suppliers competing in the market involve both big central suppliers and distributed energy provider. Meanwhile, intermediate agents containing wholesalers and retailers, energy transporters like power grid and heating company, are all regarded as equal players in the market.

Novel patterns for energy transactions are burgeoning in energy Internet, including B2B (Business-to-Business), B2C (Business-to-Customer) and C2C (Customer-to-Customer), which are typical types. B2C is similar to traditional way whereas customers in the B2C mode have more choices and rights because of the increase of energy suppliers. Companies involved in this system are motivated to provide more premium service to users. In the B2B mode, there is a strong collaboration between two companies. This cooperation can lead to a win-win situation and generate more social benefits. The last mode, C2C, is a totally new conception for energy market, which means that among the users are energy transactions. For example, a user who has surplus electricity generated by his own distributed generator may sell it to the grid or sell it to his neighbor. His choice depends on the price in the power market and the concerted price with his neighbor. The former is a kind of C2B mode, and the latter stands for C2C, a special transaction which can only happen in energy Internet.

Novel business mode of energy Internet needs a technical support. Due to the fact that traditional energy transaction only happens in a single way, the existing technologies for energy transaction are far from enough to satisfy the demands of these new ideas. Various transactions happen in this system and thousands of data are generated in a short time. The safety and reliability of these transactions are even more severe problems. A third party is needed to ensure that every transaction happening in the energy system is effective and allowable. As a result, there are many problems in this energy market model, such as low efficiency, high communication cost and over-centralization. Every single corporation will find that managing such a market is a hard problem, especially on collecting the data and guaranteeing the effectiveness.

In 2016 April, the first-ever energy transaction based on block chain happened between two residents living in Brooklyn, New York, which is called TransActive Grid and cofounded by LO3 Energy, a green energy startup, and ConsenSys, a decentralized application startup. In this project, the owner who has extra energy form PV successfully sold several kilowatts electricity to his neighbor, which is supported by block chain. As its literal meaning indicates, block chain is a chain by which numerous blocks are connected with each other. It is generally regarded as a distributed, decentralized, and public ledger. However, this analogy can be confused sometimes. We know that to make a specific transaction effective, there must be its record, which are commonly saved by the third party, like banks and so on. Let us consider how a modern transaction happens. Consumers may pay cashes to sellers to get the commodities they want. The reason this transaction could happen is that cash has credits endowed by national government. For a situation without cash like paying by Alipay, the pillar is that Alipay has its own credit. All of the people believe in Alipay thus there is a basic consensus. Block chain actually plays the same role as government or Alipay. The difference is that block chain is not a company but a big database maintained by all the participants in the market.

Block chain has been widely used in a number of fields including Internet of Things (IoT), social service, and financial system. For instance, block chain is adopted in IoT to enable the sharing of services and resources, and automate several existing time-consuming workflows in the form of cryptographic authentication [1]. As for energy trading, many researches have shown that the introduction of block chain has a positive influence on the whole system. These advantages contain reducing costs; enhancing the system security and establishing are transparent and reliable credit system, especially for the consumers.

Several countries have conducted projects on how to utilize block chain technology in energy transaction system [2]. Except for the before-mentioned TransActive Grid in America, Australia, and China have also done a lot of work. Power Ledger, founded in Perth, Australia by Ledger Assets, an Australian block-chain software company, builds a P2P surplus solar power transaction system by using block chain based-software. This system can help users who have surplus power sell this energy to other residents directly at a higher price than selling to power company. However, centralized power grid and energy suppliers won't benefit from this project at all thus the main challenges for Power Ledger will be regulatory pressure and funding. China has established CARE (Carbon Reduction Efforts Lab) lab, which develops the application of block chain technology in the sustainable development sector for green asset management.

Despite the bright future of the use of block chain in energy system, there are also many challenges [3]. Some of them are from technical aspects, while government and policy also have a strong influence. Block chain technology itself is far from mature and there is little experience of its application on energy system. The technical dilemma mainly lies in three aspects, computing capability, underdeveloped IOT technology and privacy safety risks in an asynchronous consensus network. Besides these physical problems, policy challenges are also inevitable. The decentralized energy that block chain stands for will restructure the entire ecosystem. Electricity is a crucial resource for country safety, thus its stability has to be guaranteed. Any attempts that may lead to risks should be avoided. This is first thing that should be taken into account. In the policy level, government also has to overcome the resistance from original stakeholders whose interest may be affected by block chain, break the monopoly of big cooperation and establish a management policies for novel trading systems.

With hope and challenge, block chain has begun to get on track and get attention from all over the world. The inherent consistency between it and energy Internet shows that this technology has a strong potential to be used in energy system, especially for energy transaction. Although block chain is not perfect at the monument, it is still one of the most promising technical for energy Internet.

8.3 Energy big data and cloud computing

8.3.1 Definition of big data and cloud computing

In recent decades, big data becomes a cutting-edge technology in industry and academia. "Data have penetrated into every industry and business area today and have become an important factor in production. The discovery and use of big data signal a new wave of productivity growth and consumer surplus." Said McKinsey & Company, a leading management consulting firm. Big data is a kind of novel technology that deals with large number of data including data collection and preparation, data storage and particularly, data processing and analysis, which is the core of big data. Some researchers believe that big data analytics may be a more appropriate term. This concept is widely illustrated by a set of "V"s [4], namely volume, velocity, variety, variability, veracity. Big data was initially introduced in space exploration, weather forecasting, medical genetic investigations, etc. It also works in social media like Wechat, YouTube, and so on.

There is also another definition for big data. As it literally indicates, big data is the large data sets that are output by a variety of programs. The part of data analyzation is considered in cloud computing, which refers to the mechanism that remotely takes the data in and performs any operations specified on that data. So generally speaking, big data enables us to collect and store tremendous data, and cloud computing allows us to analyze them. If we only have big data, there will be data sets that only stand alone and the value behind them will not be revealed. Likewise, if big data does not exist, then cloud computing is totally unnecessary. Since the data is not in a large scale, we can cope with it on our own computers and there is no need to send the data to "cloud." The two linked notions are perfect match and sometimes, even when we only mention on of them, we are actually talking on the two.

It is believed that big data and cloud computing have a strong compatibility with energy Internet and will become a significant support for the development of energy Internet. The core of energy Internet is applying Internet technology and clean energy-to-energy generation, energy transmission, etc. Electricity, gas, traffic, and information are all integrated and energy is efficiently shared and delivered in the whole system. To coordinate and control all the participants in energy Internet, it is important to collect, transmit, and analyze real-time data to determine a managing strategy. This raises a high and strict demand for communication techniques. In other words, to achieve "observability" and "controllability" of energy system, obtaining real-time panoramic data is essential. Based on these demands from energy Internet, big data and cloud computing become a feasible solution due to its strong data collection and analyzation ability.

8.3.2 Big data and cloud computing architecture

There are several requirements for big data and cloud computing technologies [5]. The first one lies in the processing of data. Since the amount of data generated by energy Internet is extremely large and the format is also various, integrating multiple data sources to produce more informative and synthetic information is an ideal way to deal with the huge data. For instance, to analyze demand response potential, it is necessary and convenient to consider integrated data from distribution automation systems, power quality management, and energy conservation service. The second one is data storage. As the analyzation ability of a single server is rather limited, distributed calculation is frequently used, which needs distributed databases. The stored data must be able to be visited in real time and be stretched seamlessly to a larger scale. Its reliability and rapid response are also important. The third one lies in the data analyzing algorithms. As energy Internet is a more complicated and active system than smart grid and it is influenced by severe uncertainties and various environmental factors like market policy, weather, etc., general methods have difficulties in describing such complex situations. Therefore, data driven methods that is free from physical models will be a promising solution, which include support vector machine, artificial neural network, deep learning, etc.

Lambda architecture, which was first introduced by twitter, is the model for the next generation platform of big data. It can support semi-structured, unstructured data processing, data visualization with high processing performance [6]. It has three layers, including batch layer, speed layer, and serving layer. Generally speaking, in a big data analyzation system, real-time performance and reliability of data are contradictory, which means that the faster you want to get the data, the less reliable the data will be. Lambda architecture can solve this dilemma. When the data with a large amount (PB level) reaches to this system, it will be sent to batch layer, which uses a reliable method, such as Hadoop MapReduce, to analyze the data. The generated results are saved as batch view. In this process, the data is easy to be stored while it may take a long time due to the exact calculation. Meantime, data is also sent to speed layer, which gives real-time results. Speed layer is generally based on screaming calculation platform like Storm and it uses fast incremental algorithm to get the result. This kind of calculation is rather fast but because of the inherent mechanism, data may probably get lost or the results may be wrong. Overall, batch layer uses a reliable method to calculate all the data off-line with a slow speed while the speed layer deal with the data on-line and gives a fast but less reliable result. The serving layer merges results from batch layer and speed layer and outputs the final outcome. The benefits of Lambda architecture are that it can analyze historical data and real-time data simultaneously and both real-time performance and reliability are taken into account. However, it costs more calculating resources because a single set of data is computed two times (Fig. 8.1).

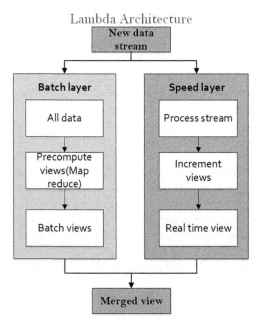

FIGURE 8.1 Lambda architecture.

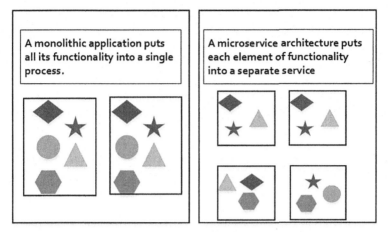

FIGURE 8.2   Difference between monolithic and micro-service architecture.

As for the cloud-computing platform, there are typically two kinds of architecture, namely monolithic and Micro-service architecture. Monolithic architecture is built as a single, unified unit. Server side application logic, front end client side logic, background jobs are all defined in the single massive code base, which means that if updates or changes are needed, developers have to build the entire stack all. This architecture is often divided into three layers, business logic layer, presentation layer and data layer. It seems to an old-dated architecture but actually in certain circumstances, it is ideal. The major advantage is that in a monolith, there are fewer cross-cutting concerns, such as logging in and rate limiting. Besides, as one packaged application, it is also less complex to deploy and has a better performance. However, it has some general problems as a single compiled system, like being hard to maintain and understand. Another one is Micro-service architecture; it is a new concept that divides the application into a series small services. Each part runs in its own process and communicates with lightweight mechanisms. It is an architecture that is better organized, decoupled, with fewer mistakes. However, it has some problems on cross-cutting concerns. The difference between the two architectures is shown in Fig. 8.2.

8.3.3   Typical application scenarios

8.3.3.1   Evaluation of policies and market mechanism

Big data and cloud computing can be used to validate the policies proposed by government and evaluate the whole energy market. Energy policies are made to guide users to participate in demand response to promote the utilization efficiency of energy system and reduce waste. What kind of policy to make, are considerably affected by energy consumption structure, energy endowment, and social and economic development. Since the energy system is complex, it is generally pretty hard to evaluate the effectiveness of energy policies. Based on historical and real-time data, big data and cloud computing can provide a reference for policy makers.

8.3.3.2 Energy production prediction

As there is more and more renewable energy utilized in energy Internet, the uncertainty of the whole energy system is continually increasing. Forecast of energy production, especially for renewable energy like solar power and wind energy, becomes an essential problem. The key to these predictions lies in the analyzation of meteorological data and historical generation data, like wind velocity, solar intensity, historical energy production, and so on. Based on the huge database, big data and cloud computing are effective to solve this problem. They can give more accurate results and promote the performances of the whole system.

8.3.3.3 Evaluation and optimization of device performance

Big data and cloud computing are ideal choices when they comes to device evaluation and optimization [7]. For a specific device, they can integrate international indexes to evaluate its performance and provide a comprehensive diagnose for the equipment. They are also able to forecast the faults of the device and make a maintenance plan for it. In accordance with the working status, big data and cloud computing can utilize the least resources to guarantee the normal operation of the device.

8.4 Internet of Things applications

To some extent, energy Internet is regarded as the combination of energy and information and communication technologies (ICT). As a general definition, energy Internet is divided into three sections: energy systems, network systems and ICT systems. One of the most promising communication system is IoT, which is composed of data acquisition, processing, transmission, and storage stages to provide a more robust and efficient communication system. It has various communication protocols and is able to deal with dissimilar data types. Hence, communications can be established in Machine-to-Machine and Human-to-Machine. As a communication technology, IoT has two main advantages in comparison with other techniques. It provides several specific communication and network structure for complex scenarios. The other one is that IoT reduces the power consumption and cost of device, leading to a more efficient utilization. In other words, when IoT technologies are used in energy system, the conception of energy Internet emerges.

The communication principles, protocols of IoT are quite complex. Novel communication technologies contain a low power wide area network (LPWAN), LTE, LTE-A, narrowband IoT as well as Zigbee and Bluetooth low energy. They are used to build long-range communication opportunities over unlicensed band. For more details, please refer to [8].

Based on communication technologies, the IoT-enabled applications that can achieve specific functions are crucial. The Internet part of energy Internet is just like information Internet, which is comprised of several layers defined by different functions, including transport and application layer, network layer and physical layer. The physical layer refers to the most basic components of energy system, such as the gas network, electric vehicle and wind turbine, which are exactly energy producers or consumers. The network layer, similar to its counterpart in information Internet, is related to data transmissions, data encryptions, and definition of data type. The application layer directly serves market participants including but not limited to individuals, families and industries, satisfying a specific demand like energy management and carbon reduction.

IoT-based applications become important when it comes to smart energy metering and management [9]. The smart metering can provide service providers and consumers for observing consuming energy values efficiently by the help of bi-lateral communication. Although traditional communication methods are still exploited in each stage of energy Internet, such as generation, transmission, and distribution, it is widely believed that the IoT system will play an important role in future communication systems. Each node in energy Internet, no matter it represents the producer or consumer, has been connected with others by the IoT applications. Any node can set up a conversation with others automatically in a fast and secure way. As a result, the whole energy Internet becomes a self-construction and self-organization system, which provides users huge flexibility.

These applications are also significant for smart cities and smart homes. With regard to cities, the application areas of the IoT, urban IoT are mainly focused on public services, such as intelligent monitoring and management of transportation systems, EMSs, healthcare services, and structural health monitoring (SHM) systems. As for homes, a smart home management system (SHMS) is comprised of a large number of smart devices that can communication with central controller. This system is also based on IoT applications enabling the devices to collect real-time data, analyze data, communicate with others, and respond to central controller.

References

[1] W. Naiyu, Z. Xiao, L. Xin, G. Zhitao, W. Longfei, D. Xiaojiang, Mohsen, When energy trading meets blockchain in electrical power system: the state of the art, Appl Sci 9 (8) (2019) 1561.

[2] Y. Cao, Energy Internet blockchain technology, in: Y. Cao (Ed.), The energy internet, Woodhead Publishing, (2019) 45–64.

[3] Muhammad BM, Jun Z, Dusit N, Kwok YL, Xin Z, Amer MYMG, Leong HK, Lei Y. Block-chain for Future Smart Grid: A Comprehensive Survey. IEEE Internet of Things Journal 2020.

[4] Y. Feng, Q. Yi, H. Rose Qingyang, Big data analytics and cloud computing in the smart grid, Smart grid communication infrastructures: big data, cloud computing, and security, Wiley-IEEE Press Hoboken, New Jersey, 2017, pp. 171–185.

[5] Z. Dongxia, C.Q. Robert, Research on big dta applications in Golbal Energy Interconnection, Global Energy Intersecction 1 (2018) 352–357.

[6] Rui F, Feng G, Rong Z, Jun H, Yi L, Lu Q. Big data and cloud computing platform for energy Internet. 2017 China International Electrical and Energy Conference (CIEEC). IEEE; 2018.

[7] J. Manar, J. Moath, B. Abdelkader, J. Yaser, A. Mahmoud, The Internet of energy: smart sensor networks and big data management for smart grid, Proc Comput Sci 56 (2015) 592–597.

[8] E.-N. Yasin Kabalci, Internet of things applications as energy internet in smart grids and smart environments, Electronics 8 (9) (2019) 972.

[9] team, D. Internet of Things energy management. Available from Digiteum: https://www.digiteum.com/internet-of-things-energy-management; 2019.

Further reading

Zhu X. (2019). Application of blockchain technology in energy internet market and transaction. IOP Conference Series: Materials Science and Engineering. vol. 592, no. 1, p. 012159. IOP Publishing.

Chapter 9

Application and potential of the artificial intelligence technology

Yu Luo[a], Yixiang Shi[b] and Ningsheng Cai[b]

[a]*National Engineering Research Center of Chemical Fertilizer Catalyst (NERC-CFC), Fuzhou University, Fujian, China;* [b]*Department of Energy and Power Engineering, Tsinghua University, Beijing, China*

9.1 Smart energy

Artificial intelligence (AI) is perhaps the fastest growing branch of modern technologies and has gained relevance to a wide range of industries. The definition of this cutting-edge technique, however, still remains vague and not precise. This conception is derived from another notion, "Natural Intelligence," which refers to the intelligence that has been existing in nature for thousands years like intelligence attributed to human and animal. Thus, the AI covers all the intelligence that does not exist in nature and is created by human beings. In a narrow aspect, it can also refer to the relevant information technologies such as machine learning, artificial neural network, deep learning, etc. Despite various definitions of AI, the core is that AI can make decisions depending on the information it receives, respond to the demand, and automatically carries out actions to achieve specific goals. Based on AI, machines are capable of having their own judgments, objectives, and actions and perform tasks like human beings, which dramatically improve the efficiency of whole society.

AI can be dated back to 1956 when a group of scientists, including John Mc-Carthy, Claude Shannon, attended Datmouth Conference and invented this new notion. Making machines work as human being seemed to be rather impossible. Followed by the creation of AI, a transient spring of AI came. Computers had been regarded as a machine that can only deal with numerical calculation, so it is amazing even that it can cope with some basic tasks used to be only solved by human beings. Researchers utilized AI to analyze some specific problems. Herbert Gelernter developed a program that can prove geometric theorems. Arthur Samuel, an American pioneer in computer gaming, proposed a Samuel Checkers-playing program, which was among the earliest self-learning programs in the world. A milestone of AI, perception neural network was also invented in this period. A perception machine can imitate the neurons and respond

Hybrid Systems and Multi-energy Networks for the Future Energy Internet.
http://dx.doi.org/10.1016/B978-0-12-819184-2.00009-2

to stimulations spontaneously. All the inputs are weighed before they are sent to a perception machine and the output will be a function of the sum of all the inputs, which is called activation function. The weight of input is not constant but variable during the process of training so that the outcome is right what is expected. It is rather a simple system compared with the ANN today, but the ideas showed in this perception machine are valuable. Training, feedback, and evaluation are definitely the core of AI.

After the short flourish, studies of AI gradually faded, which is basically because of the lack of new methods, strong calculating ability and financial support. As the funding and interest from government dropped off, a period from 1974–80 became an "AI winter." Driven by the success of expert system [1], AI later revived in the 1980s. British government started to fund it again to compete with Japan. However, from 1987 to 1993, AI faced another winter, which is due to the fact that the market of some general-purpose computers like DARPA collapsed and the funding from government ceased drastically. Research began to pick up again after this period. In 1997, the famous computer called DEEP BLUE beat a chess champion, Russian grandmaster Garry Kasparov, which is the first time that AI can defeat human beings in chess.

In 2006, Hinton proposed the artificial neural network based on deep learning, which is widely used now in various fields such as image recognition and unmanned ground vehicle. Adopting this deep learning algorithm, a team from Deep Mind trained "Alpha go" that defeated world champion of go game. Go game has always been regarded as the most complex game that human beings have ever invented. This victory of AI indicates that AI has developed to a new level. In recent years, AI is used not only in scenarios like robot communicating with people, competing with people, but also in the field of energy system, logistics system, etc. AI has definitely become a powerful tool for the development of modern technologies.

When it comes to energy sector, AI is giving rise to a new transformation leading the current system to a smart one. Although the whole energy system has been considered conservative and outdated, it is believed that this system will shift in the direction of intelligence. Recent years have witnessed a large change of energy industry in the way energy is generated, sold, and distributed. As various forms of renewable energy are involved, energy system is becoming clean and green. However, due to the inherent intermittency of renewable energy, the system tends to be more unstable and unpredictable. Besides, energy system is becoming significantly more complex because of the integration of power, gas, and heat network. Methods must be found to manage the growing electricity from renewable energy and improve the resilience of energy system. AI could be a useful tool to meet these needs [2].

In short, AI provides three main functions, forecast, optimization or management, and data analyses. Predicting the output of renewable energy could be the first use of AI. Based on the historical data and surroundings data, AI is able to forecast the working point of a wind turbine and photovoltaics. This predictive analysis can help energy companies cut costs, save power, and provide better services to users. The forecast is not limited to predicting renewable energy. Detecting the potential error and possible failures is also important application. For example, AI can predict system overloads and warn operators of potential transformer breakdowns. Several companies have started to develop relevant products. The startup VIA has developed a blockchain-powered solution called Trusted Analytics Chain that helps companies gather and analyze their data to predict system behavior. PreNav is a startup that's helping energy businesses digitize their infrastructure with the help of drones, lidar, and deep learning. Another startup, Anodot, also built a system that can forecast issues and send real-time alert to users.

The prediction function provided by AI can also be used in system modeling, particularly for the ones with high renewable energy penetration. Machine learning algorithm can help researcher establish a black-box model for sorts of energy devices based on the existing data. When it comes to engineering modeling, this method is even more efficient and accurate.

Energy optimization and management could be challenging problems for modern energy systems. Contrary to the traditional power grid, current energy system is composed of various devices related to power generation, power transmission, power to gas, and power to heat and so on. Energy storage is widely used and renewable energy starts to play an important role in the new energy blueprint. In addition to technical level, the energy trading system is also shifting in the direction of decentralization and distribution. As the system is becoming more and more complex, general managing strategies are reluctant to fulfill these demands. AI can be a useful tool to make a smart energy system because of its resilience and self-learning abilities.

Data analysis or data mining could be powerful when it comes with big data technology. Energy suppliers are dedicated to provide better service for users and to save energy, cut costs as much as possible. Millions of data generated from users and energy systems are of great value. Based on deep learning, AI is an ideal tool to deal with a large number of data. It can reveal the value behind data itself and bring energy companies with both economic benefits and deep insight of user preference.

When "Alpha go" is playing go game, the most amazing part is that it is able to play moves that human being has never expected. When AI is used in energy system, we can also expect that it will give us some impressive results. For a specific device, it may look like human beings and for a large number of generators, users, and energy storages; they may behave as a society, which is called swarm intelligence. Typical swarm intelligence means that because of the synergy of individuals, the group can be able to deal with much more difficult tasks. Every individual only has limited intelligence, perhaps just a few "if-else" programs, but their combination is highly intelligent.

9.2 Prediction for energy Internet

Solar power, hydrogen energy and wind energy are playing a more and more important role in modern energy system to decrease the carbon emissions and mitigate the environment problems. However, their inherent intermittency is regarded as a main obstacle to their wide use. In a power system, matching the load and supply is crucial to maintain the stability of the whole power grid. If the load is greater than supply, the frequency will fall and vice versa. Provided that the output of renewable energy can be forecast, it would be much easier to manage the energy system and improve its stability. In addition to renewable energy, the users' loads are also of great importance, which are immensely affected by the weather, climate, season, and also customers of consumers.

There are several methods to predict the energy supply or load. Take the building energy use, which accounts for 30% of global energy use, as an example, engineering method, AI-based method and hybrid method can be applied as prediction tools. They are classified by whether a thorough physical model for building is demanded. With regard to engineering method, it is inevitable to model the building with physical equations, consider all the relevant variables and solve the established equations. This pattern is referred to "white box" meaning that all the internal devices, structures and mechanisms should be taken into account. AI-based methods, known as "black box," however, do not rely on these physical equations and internal relations of the building. It can predict the load based on historical data and correlated variables of the building. The "gray box", namely the hybrid method, integrates both AI-based method and engineering modeling and it also needs internal information. As both "white box" and "gray box" require detailed information of and expert knowledge of the building, they become time-consuming and it is complicated to build a feasible model for buildings. Compared to them, AI-based approaches have the advantage of model simplicity, faster calculation speed, and learning capability, which have been widely applied in energy prediction.

There are four steps for AI-based prediction: data collection, data preprocessing, model training, and model testing. The first step is to gather correlated data, which may include but not be limited to surrounding temperature, air humidity, and sunlight intensity. These data must have direct or indirect relationship with outcome. For example, to consider the power output of wind turbine, it is reasonable to add wind velocity as an input parameter. However, air humidity may not be strongly related. If it is also taken into account, the accuracy and efficiency of the AI model may be reduced significantly. In a nutshell, the collected data cover meteorological data, occupancy data, and others. To train the AI model, the input data should be further handled and organized in a suitable format. These data preprocessing technologies include data transforming, data normalization, and data interpolation. The next step is model training, which is the core of the whole AI-based prediction. In an AI prediction model, there exist thousands of parameters to be confirmed. And the process of training or learning is actually the process of adjusting these parameters and get the most reasonable ones. Once all the parameters no longer change or change in a small scale, the training process comes to the end. The last step is to test or validate the trained model to test its accuracy and stability and other performance indicators. These indicators may include Root Mean Square Error (RMSE) and coefficient of determination (R^2).

The AI-based prediction can be classified in accordance with the algorithms used in the model, such as multiple linear regression (MLR), support vector regression (SVR), and artificial neural network (ANN) [3]. MLR is an approach that is able to model the linear relationship between a dependent variable and several explanatory variables. It is expand of simple linear regression where only one explanatory variable is taken into account. Assuming that variable Y_i has linear relationship with k variables from X_1 to X_k, the objective of MLR is to find the best estimation of this relationship.

The MLR model is defined as:

$$Y_i = \beta_0 + \beta_1 X_{1i} + \beta_2 X_{2i} + \cdots + \beta_k X_{ki} + u_i$$

β_i is the slope coefficients for each explanatory variable X_i and is what should be confirmed in MLR. u_i denotes the error term, which describes the parts that regression equation cannot explain. The reasonability of MLR considerably lies in several assumptions. The first one is that the error term, u_i, should be normally distributed with a mean of 0 and variance σ, that is, $u_i \sim N\left(0, \sigma^2\right)$. Then the explanatory variables should not be highly linearly correlated. Meantime, explanatory variables and error term should be also linearly independence. Needless to say, the linear relationship between dependent variable and explanatory variables are of great importance.

The main purpose is to find the best estimation of β_2 to make the MLR most precise. Specifically, provided that there are a set of random sampling data, $(Y_i, X_{1i}, \ldots, X_{ki})(i = 1 \sim n)$, we have residual sum of squares, $Q = \sum_{i=1}^{n}\left(Y_i - \hat{Y}_i\right)^2$ showing the difference of estimation result from MLR model and sampling data. $\hat{Y}_i$ is the prediction of the model, which is:

$$\hat{Y}_i = \hat{\beta}_0 + \hat{\beta}_1 X_{1i} + \hat{\beta}_2 X_{2i} + \cdots + \hat{\beta}_k X_{ki}$$

The β_i that can minimize Q should be the best choice. To get the best estimation, we need to solve the equations:

$$\begin{cases} \dfrac{\partial Q}{\partial \hat{\beta}_0} = 0 \\ \dfrac{\partial Q}{\partial \hat{\beta}_1} = 0 \\ \cdots \\ \dfrac{\partial Q}{\partial \hat{\beta}_k} = 0 \end{cases}$$

That is:

$$\sum Y_i - \sum \left(\hat{\beta}_0 + \hat{\beta}_1 X_{1i} + \cdots + \hat{\beta}_k X_{ki} \right) = 0$$
$$\sum Y_i X_{1i} - \sum \left(\hat{\beta}_0 + \hat{\beta}_1 X_{1i} + \cdots + \hat{\beta}_k X_{ki} \right) X_{1i} = 0$$
$$\sum Y_i X_{2i} - \sum \left(\hat{\beta}_0 + \hat{\beta}_1 X_{1i} + \cdots + \hat{\beta}_k X_{ki} \right) X_{2i} = 0$$
$$\vdots$$
$$\sum Y_i X_{ki} - \sum \left(\hat{\beta}_0 + \hat{\beta}_1 X_{1i} + \cdots + \hat{\beta}_k X_{ki} \right) X_{ki} = 0$$

When the model is determined, it should be estimated how much the model can explain the linear relationship. Coefficient of determination is generally used to measure the goodness of fit of the model, which is defined as follows:

$$R^2 = \frac{ESS}{TSS} = 1 - \frac{RSS}{TSS} = 1 - \frac{\sum \left(Y_i - \hat{Y}_i \right)^2}{\sum y_i^2}$$

Where ESS refers to explained sum of squares, RSS residual sum of squares and TSS total sum of squares. To some extent, ESS describes the linear part while the RSS shows that Y_i and X_i are not perfectly linearly correlated. They are defined as:

$$ESS = \sum \hat{y}_i^2 = \sum \left(\hat{Y}_i - \bar{Y}_i \right)^2$$
$$RSS = \sum e_i^2 = \sum \left(Y_i - \hat{Y}_i \right)^2$$
$$TSS = \sum y_i^2 = \sum \left(Y_i - \bar{Y}_i \right)^2$$

However, coefficient of determination always increases as more predictors are added to the MLR model even though the predictors may not be related to the outcome variable. It should be therefore modified to mitigate the influence of the number of predictors. The modified coefficient is:

$$\bar{R}^2 = 1 - \frac{\dfrac{RSS}{n-k-1}}{\dfrac{TSS}{n-i}}$$

In energy system, there are considerable factors determining the output of a machine. For example, the power output of a wind turbine can be the function of wind velocity and wind direction. It can also be affected by other factors like wind shear, which describes the drastic variation of wind vector in the vertical and horizontal direction. The MLR model is used to examine how much these

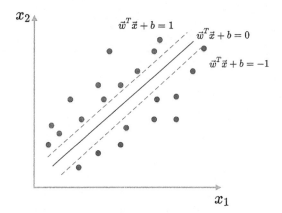

FIGURE 9.1   SVM used as classifier.

factors are related to power output and create a linear relationship between the dependent variable and all predicators.

SVR is another tool for energy output prediction, which derives from SVM (Support Vector Machine), a classifier with artificial intelligence. Fig. 9.1 shows a basic use of SVM. Assuming that there are a set of data with two features, x_i and x_2, the purpose is to find a line that can separate the blue and red ones according to their coordinates. In this situation, there can be many possible lines and SVM can help us find the best one. The best line is defined as the one that can build a gap between two categories (the red and the blue ones in this example) as wide as possible. In other words, we need to solve the following optimization problems:

$$\max_{w,b} \frac{2}{\|w\|} \quad s.t.\, y_1\left(\vec{w}^T \vec{x}_i + b\right) \geq 1, \quad i = 1, 2, \ldots, m$$

where y_i is a function of $\vec{x}_i$ that indicates the belongings of sample data. For the red category, $y_i \geq 1$ and $y_i \leq 1$ otherwise. $\frac{2}{\|w\|}$ is the geometrically the distance of the two boundary lines. It can be noted that the more the distance is, the more effective the classifier is because as far as the current sample data is concerned, the line has separated two groups as much as possible. However, only if the sample data or training data is linearly separable can we find the solution of this problem. For the example in Fig. 9.2, although it is rather simple, in this case, the dataset is linearly non-separable.

To solve this problem, we can always find a way to map the dataset to a higher dimensional feature space where the dataset will be linear separable. This approach is referred to "kernel trick." Then we can find the "line," which is also called hyperplane, in this new space.

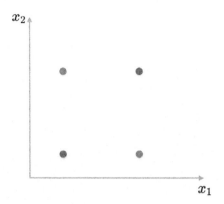

FIGURE 9.2   Linearly non-separable situation.

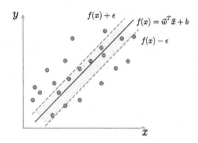

FIGURE 9.3   SVR model.

When SVM comes to prediction, we get SVR. Classical regression aims at finding a curve that can perfectly fits the sample data. The motivation is to reduce the difference between $\dot{y}$ and y as much as possible. For instance, in binary linear regression analysis, minimizing the $\sum(\hat{y}_i - y_i)^2$ is the main object. Whereas in SVR, the prediction is considered correct when the sample data fall within a small region near the regression curve. Outside this region, we will consider penalty in the objective function.

Fig. 9.3 illustrates the basic ideas of SVR model. To get the regression curve, we need to solve the optimization problem:

$$\min_{w,b} \frac{1}{2}\|\vec{w}\|^2 + C\sum_{i=1}^{m} \ell_t \left(f(\vec{x}_i) - y_i \right)$$

where ℓ is the penalty function, defined as follows:

$$\ell_\epsilon(z) = \begin{cases} 0, & \text{if } |z| \le \epsilon \\ |z| - \epsilon, & \text{otherwise} \end{cases}$$

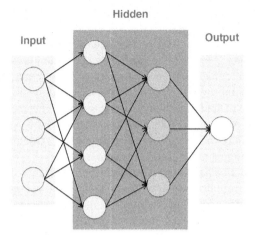

FIGURE 9.4   Structure of typical ANN.

which means that if the error is acceptable, penalty does not exist. Otherwise, the penalty is positively related to the deviation degree. C is the penalty factor, which represents how much the deviation is valued.

ANN is currently a cutting-edge technology for prediction [4,5,6]. Motivated by imitating natural neural network, ANN uses the same process to learn from training data, which is now widely used in pattern recognition. Fig. 9.4 describes a typical structure of ANN with three inputs and one output and two hidden layers. It can be simply regarded as a "black box" or a function between output and input. In this case, the training data is in the form of $x_{11}, x_{12}, x_{13}, y_1$ indicating that there are three inputs. Every point in ANN is an artificial neuron which is usually a math function imitating natural neurons, like sigmoid function. The signal entering neuron is mapped to a limited output signal. Although a neuron function is rather simple, when thousands of neurons come together, the ANN can cope with complex tasks. The link between the neurons in different layers has specific parameters, which are exactly the weight of different inputs. That is, the input is first weighed and then transferred to the neuron function. In short, when we are training an ANN model, we are actually finding the most suitable parameters for all the neuron connection. For example, given a point of data, $(x_{11}, x_{12}, x_{13}, y_1)$, we can get output $\dot{y}_1$. Obviously this estimation is different from y_1 and we have the error $\dot{y}_1 - y_1$. The next step will be to adjust the parameters in the direction of decreasing the error. By doing this, we get new parameters for ANN model and can move to the next data point. The learning process of ANN is the process of its parameter adjustment. Gradient Descent Method is an effective way to train ANN model.

Above mentioned are basic ideas of how to use artificial intelligence in the field of prediction. These ideas have been widely studied in academia and also gradually used in energy sector. In 2018, Deep Mind and Google started to use artificial neural network and machine learning in the field of wind energy. They have carried out a prediction project for 700 megawatts of wind energy capacity in the central United States. They have succeeded predicting the wind power output 36 hours later. Although the prediction is far from perfect, it could help managers make an optimal delivery commitment in advance. So far, the machine learning has improved the value of their wind energy of 20% [7]. In 2014, Google also began to apply machine learning to optimize the cooling system of their data centers. Meanwhile, they used smart temperature, lighting, and cooling controls to reduce the energy consumption. Today, Google's data center is nearly twice as efficient as a general data center [8].

AI-based prediction has shown a great potential in energy system. It requires less detailed physical information of the system, which makes modeling much easier. Only historical data and environment data are needed when modeling, which makes the prediction much more convenient. However, the decoupling of models and physical mechanism also makes it impossible to extrapolate system performance once the design and operation has changed. Also, in the design process when we have no historical data, these intelligent predictions may not be put to good use.

9.3 Control and optimization based on artificial algorithm

Traditional control methods are based on classical control and modern control theory, both of which require an explicit mathematical description of the controlled object, in the form of transfer function or state-space equation. These approaches have been of great importance in the control of conventional power and thermal system. However, these methods highly rely on the mathematical model of controlled object, either in the form of function or equation group. What makes these not easy to be used is the growing complexity of modern energy system. It is believed that building a perfect model for a system, even for a simple small energy network, would be extremely challenging. That is why we need artificial intelligence to deal with this task.

Control approach based on artificial intelligence, also known as intelligent control, is a promising direction of modern control strategy. It consists of fuzzy control, ANN control, and so on. The common ground of these control methods is the imitation of natural intelligence. They adopt different ideas to simulate the process how human beings make control of a system.

Fuzzy control is a method based on fuzzy math. Its aim is to "translate" the human prior knowledge into a form that can be understood by computer using the theory of fuzzy set and fuzzy relation. Human beings have accumulated a large number of knowledge and experience. For example, if the room temperature is too high, we might need to improve the power of air conditioner. We can also say that if the temperature is extremely higher than what expected, we need to improve the power of air conditioner as much as possible. When it comes to solid oxide fuel cell, if the current increases, the hydrogen flow rate should also

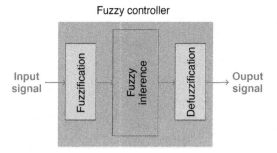

FIGURE 9.5   Fuzzy controller.

be improved. These rules can easily be understood by engineers or scientists, however, they are not able to be implemented by computers.

Fig. 9.5 shows a typical process of fuzzy control, which covers several steps, mainly fuzzification, fuzzy inference, and defuzzification. The continuous input signal, generally the system error or the error variation, which is collected from real physical terminal, is first mapped to a discrete one by using quantization factor. The discrete signals are further mapped to fuzzy sets by fuzzification method. These fuzzy sets qualitatively describe the input signal in the form of NB (negative big), NS (negative small), Z (Zero), PS (positive small), and PB (positive big). By fuzzy inference, which is actually a calculation between fuzzy set and fuzzy relation, computer can acquire the output fuzzy set, in the same form with the input sets. The next step is the inverse process of fuzzification. Output fuzzy sets are mapped to a series of discrete data and further to the control signal that can be detected by the real system. Fuzzy control is widely implemented in hybrid system. Such applications cover wind energy conversion system, battery management system, heating energy supply system, and energy storage system.

The main idea of control method based on ANN is to replace the physical model of the controlled object with its ANN model. It is known to us that ANN model is a "black box" where given an input; the output will be calculated without solving physical equations. It uses the historical data to train the model to get the most suitable parameters, as mentioned before. ANN control can be utilized when the physical mechanism is too complex to be modeled.

The optimization of energy system is another important topic, especially when more and more renewable energy and distributed subsystems are integrated. The optimization involves how to improve the efficiency of the system, how to maximize the profits and how to minimize the costs. Some of them are in accordance with economy and some are regarding physical performance. Different from general optimization methods that implement optimization algorithm to find the optimal solution of problem, due to the complexity of the system, intelligent algorithm relies on heuristic algorithm that is focused on finding a feasible solution. Heuristic algorithm derives from intuitive thinking

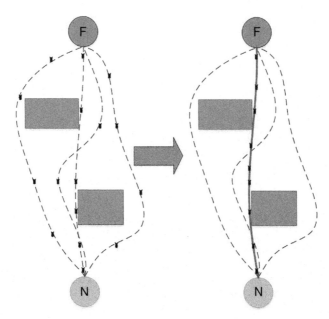

FIGURE 9.6 Foraging activities of ant colony.

and experience, which may try to imitate some natural phenomenon like the evolutionary of creatures, the migration of a bird flock. There are several types of heuristic algorithms including genetic algorithm, ant algorithm, particle swarm optimization, simulated anneal algorithm, etc [9].

In the ant algorithm, the idea of finding the optimal solution is based on foraging activities of ants. Provided that there is a food near ants' nest and between them are some obstacles, the ant colony will find a feasible and even the shortest path by self-organization, as shown in Fig. 9.6. This intelligent behavior lies in two factors. The first one is whenever an individual ant passes through a certain path; pheromones will be remained on this path. Assuming that the speed of ant is constant, it can be inferred that the shorter the path connecting food and nest is, the more pheromones will exist on this path. The second rule is that ants tend to follow a path that has more pheromones. It is also possible that an ant will go to a path with less pheromones. However, the probability could be significantly low. These two rules actually form a positive-feedback cycle, which means the more pheromones a path has, the more ants pass through this path, vice versa. Eventually, the paths of ant will converge to a specific path that may be the shortest one with a high probability.

Ant algorithm can solve the NP problems in combinatorial optimization and has been used in a variety of fields. When it is used in energy system, there appear a few valuable researches. Using multi-layer ant colony optimization, Mousa Marzbanda etc. figure out the optimum operation of micro-sources for decreasing the electricity production cost by hourly day-ahead and real time

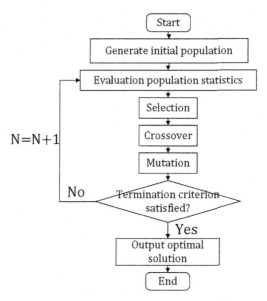

FIGURE 9.7   Flowchart for genetic algorithm.

scheduling. On mass rapid transit system, Bwo-Ren Ke, etc. propose a method for block-layout design to save energy. They used a Max-Min Ant System of ant colony algorithm, which can also reduce the computational time [10].

Genetic algorithm, also widely used, imitates the process of evolutionary of creatures [11]. It is well known that creatures are evolving to adapt to the environment by inheritance and variation. The individuals with high adaptation tend to be alive during the natural selection while the ones with bad traits will be eliminated. Fig. 9.7 gives the flowchart of genetic algorithm. The first step is to map the possible value of variables in continuous space to a chromosome. For example, assuming that the independent variable x_1 and x_2 are both in the range (0, 1), we can map them to a binary string in the form of 0010101010001 010. The first eight bits represent x_1 and the latter bits represent x_2. It can be seen that the higher the precision is, the more bits are needed to discretize independent variable space. The objective function of optimization is further used as a criterion of adaptation for the individuals from genetic model. A higher adaption means that this individual has a larger probability to be chosen to crossover with other individuals and its gene is more probably inherited. Mutation of genes in chromosome, which can be considered as a change of a bit in the binary string, is also crucial because the initial population is selected stochastically thus without mutation, the model may jump into local optimization quickly. Choosing a suitable possibility for mutation plays a major part in the calculation. Too large possibility means more iterations, slower computational speed, and bad convergence performance. On the other hand, small possibility leads to local optimization.

A research from Jonathan Reynolds, which is focused on district heat supply system, utilized genetic algorithm to manage. The authors first built a system model with ANN method and then got the optimal operating schedules of the heat generation equipment, thermal storage, and the heating set point temperature using genetic algorithm. Abbassi Abdelkader used a multi-objective genetic algorithm to optimize the sizing of a stand-alone PV and wind power supply system, where all storage dynamics are considered. The objective is to minimize the total cost electricity and the loss of power supply probability of the load. Results demonstrated renewable energy play an important role in promoting the energy sector.

9.4 Swarm intelligence for complex energy networks

Swarm intelligence (SI) is originally a phenomenon in nature and now is also achieved in artificial system. When an ant colony is taken into account, every individual ant only has extremely limited intelligence like walking in a path with more pheromones but the whole colony shows an intelligent way to find food or face disasters. Swarm intelligence is defined as a collective behavior of a decentralized or self-organized system. These systems consist of numerous individuals with limited intelligence interacting with each other based on simple principles. Although the individual is not so smart and there is no central leader dictating how individuals behave, only by the interaction between individuals can the whole system achieve a holistic intelligence. In general, no individual has an overall cognition but just carries out its simple actions.

Typically, to establish a SI system, we need to design the behavior of individual or agent and the interaction principle between agents and between agent and environment. Nevertheless, it could be hard to predict whether the system will have a SI or how intelligent the system is. The system will not be able to show intelligence at one time but behave beyond our expectation at another time. When the SI emerges, it is called swarm intelligent emergence.

On the blueprint of energy network, every participant is able to make its own decision in energy market and respond to environment change and demand from other participants. As a result, it is impossible to manage all the families, factories, and companies. All of them become intelligent individual, and with suitable design, we may see a SI emergence. Provided that there is no SI emergence, it is also possible to achieve it by improving the intelligence of individuals through self-learning. The system becomes more effective and the individual gets smarter. There is a positive feedback between system and individual just like our society, where people are promoting themselves to make a better society and an ideal society will provide people more opportunities of self-promotion. This process can be called swarm intelligent evolution.

Alphago-zero provides a good example of self-learning. Compared to alphago, the famous program beating Lee Sedol, alphago-zero is able to learn from zero. It does not use any data from human beings but only plays games against itself to achieve self-promotion. From stochastic play, using reinforcement learning algorithm, it became more and more powerful. After 3 days' training, it was even stronger than any previous go game programs and also human beings.

Alphago-zero demonstrates that the individual can improve its intelligence by self-learning. The algorithm cannot only be used in go game but also other fields. It can be predicted that the devices in the future energy system can learn from both the environment and itself to achieve individual intelligence. Further, the whole energy network appears swarm intelligence emergence and generates more data that provides individuals with new learning environment. In this way, the entire energy network will become immensely efficient and clean.

Although there is still a long way to go to acquire swarm intelligence, there have been many related and valuable researches. Most of them are focusing on how to improve the efficiency of a specific energy system, how to reduce energy cost, how to optimize design based on all sorts of intelligent methods, like deep reinforcement learning and heuristic algorithms. These researches are mostly regarding building individual intelligence, which plays a crucial role in swarm intelligence.

A good example of swarm intelligence comes from one of our latest researches on energy network [12]. Given a small energy network consisting of eight subsystems with similar architecture, the objective is to achieve a swarm intelligence using multi agent system algorithm. Every subsystem is given an intelligence to adjust its power output according to its surroundings and the information it receives. Between every two points are information links while the physical connections, mainly power links, are limited to certain edges. To evaluate the collective intelligence, we assume that there is a sudden load change in subsystem Num. Seven and then observes the autonomous respond of the whole network. Furthermore, the influence of physical topology on this respond is discussed.

To be simple, every subsystem in this network can be viewed as a small energy system where there are both energy production and consumption. The whole energy utilization rate of this system highly relies on its energy output. Fig. 9.8 shows five possible and typical topologies of the eight systems and the edges in these graphs denote power connections. As a network, to sustain system stability, it is important to balance the real-time power output and consumption of the whole network. Therefore, we have an optimization problem:

$$max\,\eta all\,(p1, p2, \cdots, p8) = \frac{\sum_{i=1}^{8} pi}{\sum_{i=1}^{8} \dfrac{pi}{\eta_i(i)}}$$

$$s.b.\sum_{i=1}^{8} pi = Demand$$

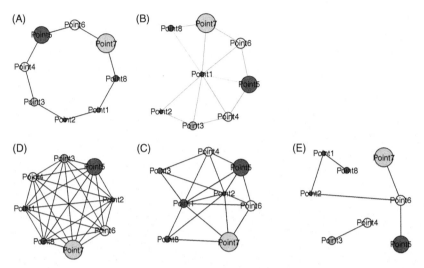

FIGURE 9.8 Alternative topologies for the energy network.

where the power loss in transmission is ignored. To determine every power output, it is without doubt effective to solve this optimization problem using conventional mathematics approaches and then control every energy system by central control unit. However, a decentralized intelligent algorithm could be more applicable and attractive when it comes to real time control. To achieve swarm intelligence, we simply gives every system an agent to make independent decision, followed by building a specific communication protocols between every two points. The protocols can be expressed as:

$$p_i^{k+1} = p_i^k + \Delta p \sum_{j \in N_i} \mathrm{sgn}\left(\frac{dc_j\left(p_j^k\right)}{dp_j^k} - \frac{dc_i\left(p_i^k\right)}{dp_i^k} \right)$$

This equation denotes the strategies based on which every system changes its output. Each system will consider and only consider all the nodes that are linked with it to make its own decision. Fig. 9.9 illustrates how every system respond in the presence of a sudden load change of system 7. It can be seen that at the start, the output of system 7 increases steeply to satisfy its own demand. However, this load is significantly bigger than its rated output and results in an efficiency decrease of the whole network. It is shown that other system starts to increase its output and transmit extra energy to system 7. How much energy will be transmitted depends on each system's rated output power. Fig. 9.10 further shows the efficiency change in this process where different topologies are also taken into account. The efficiency first drops notably and then increases to a high level, which even surpasses the efficiency at the time before the load change. It is also illustrated that the speed of network autonomous coordination is deferent when topologies are allowed for. When all the systems are connected

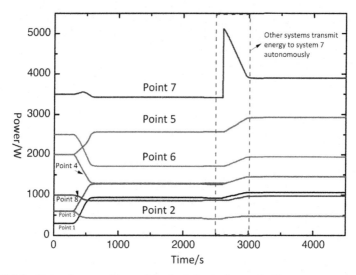

FIGURE 9.9 Network respond to a sudden load change from point 7.

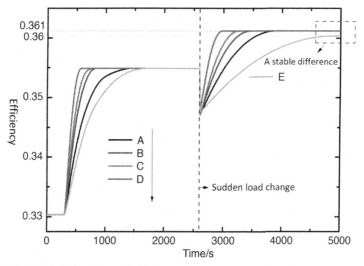

FIGURE 9.10 Variation of overall efficiency when load changes abruptly.

and energy transmission is allowed between every two points, the network has a rapid response, that is to say the network is "smarter." If there are only few connections, the network becomes dull. The graphical analyzation will give more useful information of this topic.

Although the above-mentioned example is rather simple and more likely a decentralized control, it is more or less helpful to illustrate the conception, swarm intelligence. In this example, the ability of self-decision can be viewed as an individual intelligence and the collective behavior of the network when it faces a sudden load change is a swarm intelligence emergence. What is interesting is that we cannot predict the power output change of these energy systems. They are not controlled directly by human beings but they are able to optimize themselves by self-organization, which is a good example of their intelligence.

It is believed that with more and more AI related studies in energy sector; the achievement of swarm intelligence is not impossible. We hope a better and more intelligent energy network.

References

[1] C.N. Ernesto, T.P. Mario, J.C. Juan Carlos, C.S. Roberto Valentín, R.M. José Gabriel, Expert control systems for maximum power point tracking in a wind turbine with PMSG: state of the art, Appl Sci 9 (12) (2019) 2469.

[2] C. Donitzky, O. Roos, S. Sauty, A digital energy network: the internet of things and the smart grid, Intel (2020).

[3] W. Zeyu, S. Ravi S., A review of artificial intelligence based building energy use prediction: contrasting the capabilities of single and ensemble prediction models, Renew Sustain Energy Rev 75 (2017) 796–808.

[4] J.-L. Casteleiro-Roca, Prediction of the energy demand of a hotel using an artificial intelligence-based model, Hybrid Artificial Intelligent Systems, Springer, Cham, 2018, pp. 586–596.

[5] F.B. Jesús, F. Juan, F. Gómez, P. Fernando Olivencia, M. Adolfo Crespo, A review of the use of artificial neural network models for energy and reliability prediction. a study of the solar pv, hydraulic and wind energy sources, Appl Sci 9 (2019) 1844.

[6] R. Jonathan, W.A. Muhammad, R. Yacine, H. Jean-Laurent, Operational supply and demand optimisation of a multi-vector district energy system using artificial neural networks and a genetic algorithm, Appl Energy 235 (2019) 699–713.

[7] Witherspoon S, Fadrhonc W. Machine learning can boost the value of wind energy. Avialble from Google: https://www.blog.google/technology/ai/machine-learning-can-boost-value-wind-energy/; 2019.

[8] Hölzle, U. Data centers are more energy efficient than ever, Avialble from Google: https://www.blog.google/outreach-initiatives/sustainability/data-centers-energy-efficient/; 2020.

[9] M. Mousa, Y. Ebrahim, S. Andreas, D.G. José Luis, Real time experimental implementation of optimum energy management system in standalone microgrid by using multi-layer ant colony optimization, Int J Electr Power Energy Syst 75 (2016) 265–274.

[10] K. Bwo-Ren, I.E.E.E. Member, C. Meng-Chieh, L. Chun-Liang, I.E.E.E. Senior Member, Block-Layout Design Using MAX–MIN Ant System for Saving Energy on Mass Rapid Transit Systems, IEEE Transact Intell Transp Syst 10 (2009) 226–235.

[11] A.R. Abbassi Abdelkader, Multi-objective genetic algorithm based sizing optimization of a stand-alone wind/PV power supply system with enhanced battery/supercapacitor hybrid energy storage, Energy 163 (2018) 351–363.

[12] Z. Yi, S. Yixiang, L. Yu, G. Zhongxue, Coordination control for distributed energy based on multi-agent system, Chin Sci Bull 62 (2017) 3711–3718.

Index

Note: Page numbers followed by "f" indicate figures and "t" indicate tables.